No. 1412
$13.95

The Backyard Energy Projects Handbook

by William M. Foster

TAB BOOKS Inc.
BLUE RIDGE SUMMIT, PA. 17214

FIRST EDITION

FIRST PRINTING

Printed in the United States of America

Library of Congress Cataloging in Publication Data

Foster, William Murchison, 1931-
The backyard energy projects handbook.

Includes index.
1. Solar energy. 2. Wind power. I. Title.
TJ810.F67 621.47 82-5681
ISBN 0-8306-2512-7 AACR2
ISBN 0-8306-1412-5 (pbk.)

Contents

Other TAB books by the author:

No. 1006 *Build-It Book of Solar Heating Projects*

Introduction

With the cost of oil and electricity continuing to spiral upward, the sun has become a viable source of energy. It is no longer viewed with skepticism nor as a futuristic commodity or scientific notion shrouded in mystique. "Will it work?" has been replaced by "How can solar energy work for me?"

Major changes in our energy usage will develop through legislation, tax incentives, and pricing policies relatively independent of our individual control. These changes will involve some dislocations in life styles, but perhaps there is a way to save energy and have some enjoyment in the process.

Have you ever tried cooking on a solar barbecue or raising vegetables or flowers in a solar-heated greenhouse? These and other alternative energy sources merit your consideration, especially now that the cost of travel encourages more of us to spend leisure time at home.

The *Backyard Energy Projects Handbook* will provide you with ideas for alternative energy use. They are all practical, fun, and educational, and will interest and involve the whole family.

1

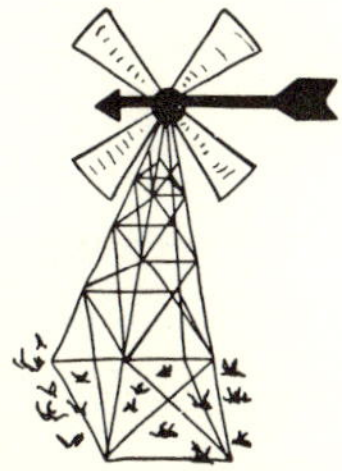

Solar Cooking and Food Drying

The sun has been used worldwide since ancient times to preserve and prepare food. Artwork dating back to 1900 B.C. shows Egyptians using the sun to dry fish and meat. As early as 1860, soldiers in the field received rations cooked by the sun. Since that time, most interest in this technology has been promoted by necessity in underdeveloped countries who have arid or tropical climates and lack the resources to cook by our standards. India, for example, has been a prime mover in the design of solar cooking devices. Fuel is so scarce there that dried cow dung is traditionally burned to cook meals. This practice in turn deprives an arid land of the fertilizer required for agricultural development.

Many of the designs in this chapter are a result of studies sponsored by the Ford Foundation, Volunteers in Technical Assistance (VITA), the United Nations, and universities around the world to assist countries such as India, Mexico, Egypt, and Pakistan.

THE SOLAR STOVE

For Americans, a solar stove provides a practical means of outdoor cooking in remote locations and an enjoyable placement to the backyard barbecue. If the weather is too overcast to cook with the sun, you may need to fire up that smoky charcoal burner, though, so don't discard it!

In selecting solar cooking devices, consider both climate and the style of cooking you most enjoy. Hazy sunshine, intermittent clouds, and wind all reduce the effectiveness of the unit. Also, in the morning and late afternoon, low sun angles provide less heat, so breakfast and dinner may be the most difficult meals to prepare. A solar lunch, on the other hand, can be most rewarding.

Portability is desirable for many families using a solar cooker for picnics, camping trips, or other outings. In Europe, skiers and mountain climbers have effectively used lightweight cookers for many years. Portable cookers must not be flimsy, however. You can imagine the frustration of a solar chef whose meal has been upended by a gust of wind!

TYPES OF SOLAR COOKERS

Three types of solar cookers have become most popular with today's sun-powered chefs. These provide the best efficiency and ease of operation and can be easily adapted to fit almost any need.

Concentrators

A concentrator solar cooker is one that uses curved reflective surfaces to focus energy onto the food or cooking vessel. Although you can also focus sunlight with a lense, such as a magnifying glass (most of us have burned holes in leaves or other objects this way), but only *reflective* surfaces are considered here because they have been found to be the most practical.

To understand the concept, picture an opened umbrella turned upside down with the handle pointing toward the sun. If the umbrella were constructed of reflective material and the food or pot containing it were supported by the handle at the proper distance from the reflector, the result would be a concentrating-type solar cooker. The technology is a bit more complicated than this, since the reflector is normally curved in the form of a parabola (see Fig. 1-8). The food is placed at the focal point of the parabola, the point at which sunlight is most intensely focused. (A parabola is the shape formed by cutting a right circular cone parallel to either side of the cone.) An umbrella-type cooker is illustrated in Fig. 1-1. As shown, some type of support structure, such as a tripod, is needed to hold the umbrellalike part. You also must be able to adjust the reflector to follow the sun as it travels across the sky while keeping the food grill fairly level in the process. This kind of unit was sold in the United States in the 1960s under the trade name Umbroiler.

Another form of concentrator bends the reflective material into the shape of a parabola forming a troughlike structure that can be adjusted to track the sun along one axis only. These are also called *line focusing* concentrators, because the energy is concentrated over the entire length of the trough rather than at a specific point as with the umbrella cooker. If you place a spit or "cooking tube" at the focus line, you can cook hot dogs easily.

A well-designed concentrator cooker starts collecting usable heat as soon as direct sunlight is focused on it. It is possible, therefore, to collect heat at almost any time of the day—even for an early morning breakfast. Temperatures approaching that of an open fire are possible with a good reflector, strong clear sunlight, and the absence of wind. Cooking times for eggs, steak, or fish are approximately equivalent to a barbecue.

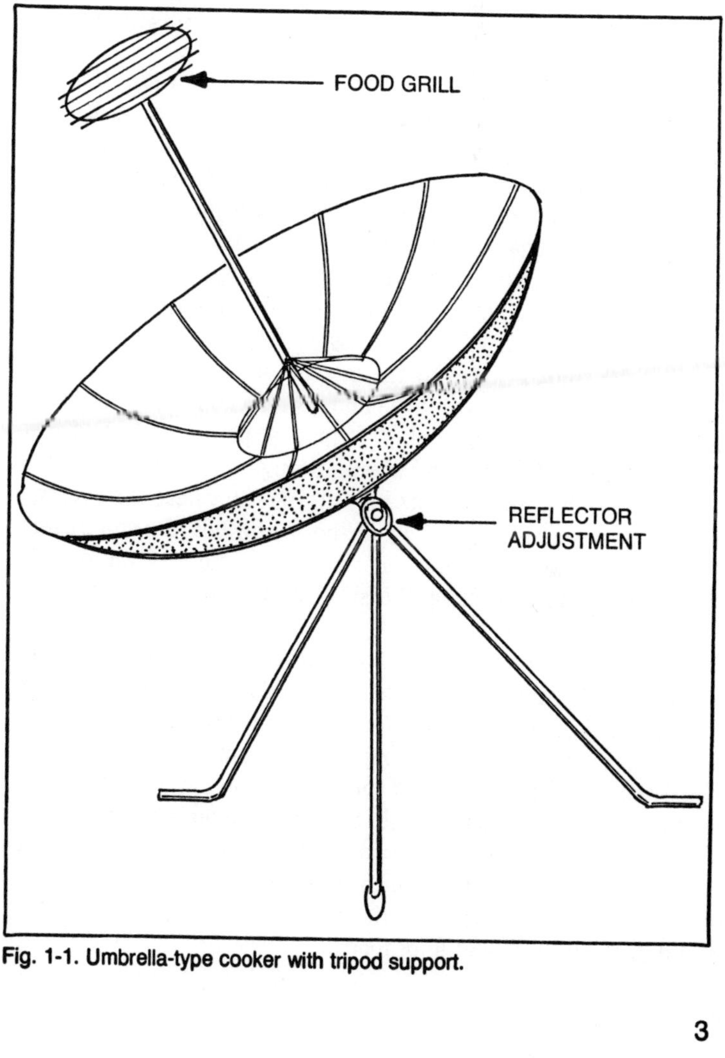

Fig. 1-1. Umbrella-type cooker with tripod support.

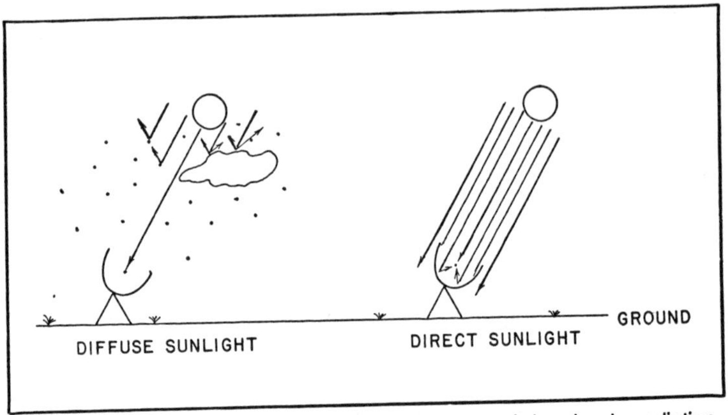

Fig. 1-2. Clouds and air contaminants reflect, scatter, and absorb solar radiation.

To operate effectively, a concentrator must have *direct,* bright sunshine and be moved to track the sun as it travels across the sky. With these requirements, the focusing cooker is best suited to those areas that have a high ratio of direct-to-diffuse sunlight. The concentrating collector operates best with clear skies and poorly on cloudy or overcast days. Clouds, dust, smoke, and smog in the air all tend to reflect, scatter, and absorb solar energy (Fig. 1-2). This *scatter effect* is most noticeable in large cities where air pollution diminishes the amount of direct sunlight. Table 1-1 ranks some key cities in the United States by the amount of mean sky cover from sunrise to sunset. The higher the number, the more difficult it is to use a concentrating-type cooker. It may surprise you to find that Hilo, Hawaii, with an annual average temperature of 73° F, has almost as much cloud cover as Juneau, Alaska, where the average temperature is a mere 40° F. In northern latitudes, the sun travels at lower altitude angles and must penetrate more of the earth's atmosphere before reaching the surface. This reduces the direct radiation available (Fig. 1-3).

In summary, a concentrator will produce the highest cooking temperatures of any solar cooker, provided it is operated in direct sunlight and focused frequently (every 10 to 15 minutes) to keep the maximum energy on the cooking pot. Concentrators collect heat rapidly but also lose it quickly if wind is present. Large quantities of food are difficult to cook in this manner. If liquids are involved, frequent stirring is necessary to prevent scorching. Cooking with a concentrator may be compared to cooking over a gas flame or the burner of an electric range on the high heat setting. It is best for frying, broiling, and boiling of foods.

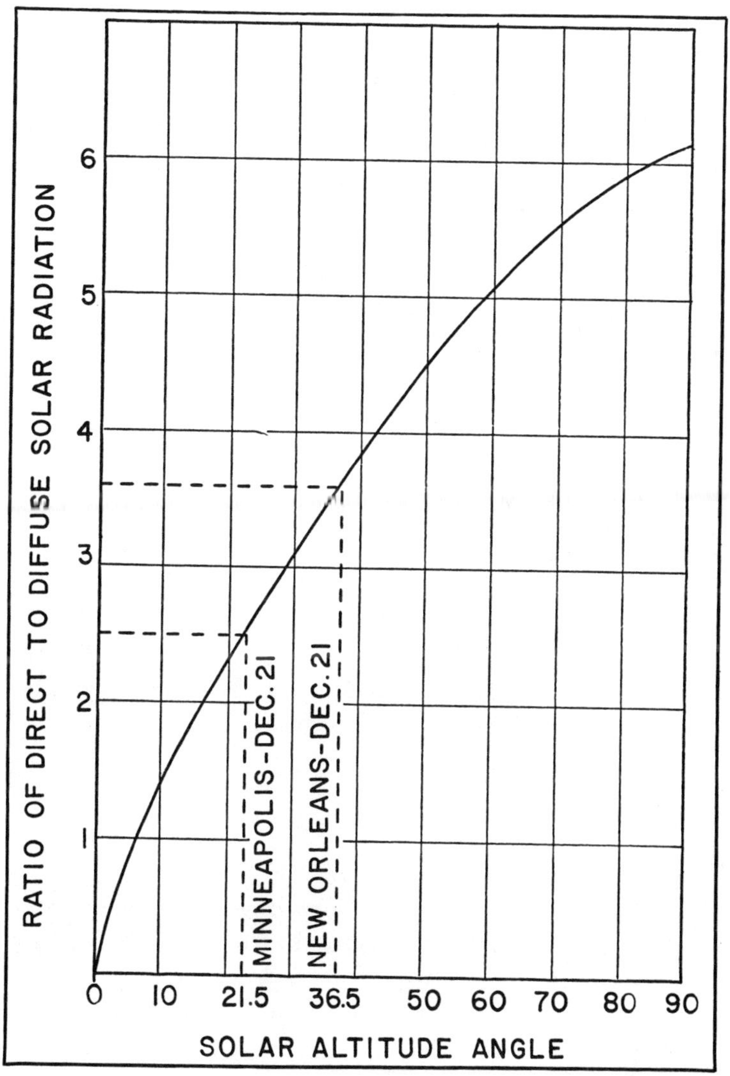

Fig. 1-3. Comparison of atmospheric effect created by sun angles at noon on a cloudless winter day in northern (Minneapolis, latitude 48°) and southern (New Orleans, latitude 30°) locations

Table 1-1. Mean Sky Cover for Selected Cities, Sunrise to Sunset (10 Represents Complete Sky Cover).

Location		by Month												Annual
		J	F	M	A	M	J	J	A	S	O	N	D	
Alabama	Birmingham	6.2	5.9	5.6	5.2	5.2	5.2	5.7	5.3	4.8	4.1	4.7	6.1	5.3
	Mobile	6.5	6.6	6.0	5.7	5.6	5.7	6.8	5.4	5.8	4.3	4.9	6.1	5.8
	Montgomery	5.9	5.7	5.2	4.8	4.7	5.1	5.5	5.2	4.5	3.9	4.5	5.7	5.1
Alaska	Anchorage	6.8	6.8	6.6	6.9	7.6	7.5	7.6	7.8	7.8	7.6	7.3	7.4	7.3
	Annette	7.7	8.0	7.7	7.6	7	8.0	8.0	7.8	7.7	8.6	8.5	8.6	8.0
	Barrow	*	5.3	5.0	6.0	8.5	8.0	8.2	8.9	9.2	8.8	*	*	-
	Barter Island	*	4.7	5.8	6.5	8.5	8.1	7.9	8.3	8.6	8.5	*	*	-
	Bethel	6.3	6.2	6.5	6.9	7.9	8.1	8.5	9.0	8.2	7.7	7.2	6.7	7.4
	Cordova	7.2	7.1	7.1	7.4	8.5	8.3	8.6	8.1	8.2	8.0	7.9	7.8	7.9
	Fairbanks	6.5	6.3	6.2	6.2	7.0	7.3	7.3	7.8	7.9	7.9	7.1	7.1	7.1
	Juneau	7.9	8.0	7.8	8.2	7.8	7.8	8.3	8.0	8.4	8.6	8.5	8.5	8.2
	Kotzebue	5.6	5.5	6.0	5.8	6.4	6.9	7.7	8.3	7.5	7.2	6.6	5.9	6.6
	Mc Grath	6.6	6.3	6.5	6.1	7.5	7.8	7.8	8.5	8.1	8.4	7.4	6.8	7.3
	Nome	5.9	5.8	5.9	6.5	6.7	6.6	7.9	8.1	7.5	7.1	6.5	6.6	6.8
	St. Paul Island	8.1	8.0	7.7	8.1	8.7	8.9	9.1	9.1	8.6	8.5	8.3	8.4	8.5
Arizona	Flagstaff	5.9	4.9	5.0	4.9	4.1	2.4	5.4	5.3	2.9	3.1	3.4	4.4	4.3
	Phoenix	5.1	4.5	4.3	3.8	2.9	2.0	3.8	3.5	2.0	2.8	3.1	3.6	3.5
	Tucson	5.0	4.5	4.6	3.7	2.8	2.2	5.5	4.7	2.7	3.1	3.0	4.4	3.9
	Yuma	2.4	2.5	2.2	1.6	1.1	0.6	1.7	1.8	1.1	1.3	1.7	2.5	1.7
Arkansas	Ft. Smith	5.4	5.5	5.4	5.4	5.2	4.6	4.4	4.2	3.9	3.3	4.5	5.3	4.8
	Little Rock	6.2	6.3	6.5	6.4	6.2	5.7	6.0	5.4	4.8	3.9	5.0	5.9	5.7
California	Bakersfield	6.2	5.5	5.4	4.4	3.2	1.3	1.1	1.1	1.3	2.7	4.1	6.3	3.6
	Bishop	5.3	4.8	4.7	4.7	4.1	2.3	2.4	1.9	2.0	2.8	3.5	4.9	3.6
	Eureka	7.3	7.3	7.3	7.2	6.8	6.4	6.6	6.9	6.0	6.5	7.0	7.6	6.9
	Fresno	6.7	6.1	5.3	4.4	3.4	1.9	1.1	1.1	1.5	2.8	4.4	7.0	3.8
	Los Angeles	4.6	4.8	4.8	5.3	4.8	4.1	2.8	2.7	2.8	3.9	3.4	4.5	4.0
	Red Bluff	6.6	6.3	6.1	5.4	4.6	3.2	1.3	1.6	2.2	4.0	5.5	6.9	4.5
	Sacramento	6.0	5.2	4.4	3.5	2.7	1.6	0.7	0.7	1.4	2.4	4.0	5.9	3.2
	San Diego	5.1	5.0	4.9	5.7	5.5	5.1	4.4	4.3	3.7	4.3	3.7	4.7	4.7
	San Francisco	5.7	5.5	5.0	4.5	4.4	3.7	4.2	4.6	3.8	4.0	4.5	5.6	4.6
Colorado	Colorado Springs	4.9	5.1	5.7	6.0	6.0	4.5	4.9	4.7	3.5	3.7	4.7	4.5	4.9
	Denver	4.5	4.8	5.3	5.7	5.7	4.7	4.6	4.7	4.0	4.1	4.2	4.4	4.7
	Grand Junction	5.3	5.4	5.3	5.4	4.8	3.6	3.9	4.0	3.3	3.5	4.2	5.1	4.5
Connecticut	Hartford	6.2	5.7	5.8	6.1	6.0	5.8	5.7	5.6	5.4	5.3	6.2	6.2	5.8
District of Columbia	Washington	6.1	5.6	5.6	5.4	5.4	5.1	5.1	5.1	4.8	4.7	5.4	5.8	5.3
Florida	Apalachicola	5.2	5.4	5.5	4.8	4.3	5.2	6.0	5.7	5.6	3.9	4.4	5.7	5.1
	Jacksonville	5.3	5.2	4.9	4.6	4.7	5.6	5.9	5.6	5.7	5.1	4.7	5.4	5.2

	Key West	**4.3**	**3.8**	**3.7**	**3.8**	**4.4**	**5.4**	**5.3**	**5.3**	**5.5**	**4.9**	**4.2**	**4.3**	**4.6**
	Miami	5.0	4.6	4.7	4.9	5.4	6.2	6.1	5.8	6.2	5.8	5.2	5.2	5.4
	Tampa	5.3	4.7	4.8	4.7	4.6	5.8	6.3	6.1	6.0	4.7	4.6	5.2	5.2
Georgia	Atlanta	[illegible].3	6.2	6.1	5.5	5.5	5.8	6.3	5.7	5.3	4.5	5.1	6.2	5.7
Hawaii	Hilo	[illegible].8	7.0	7.8	8.3	8.2	7.6	7.8	7.9	7.2	7.4	7.6	7.6	7.6
	Honolulu	[illegible].3	6.0	5.9	6.0	6.0	5.5	5.0	5.1	4.8	5.4	5.5	5.4	5.5
	Lihue	6.2	6.5	6.5	6.9	7.0	6.6	6.5	6.6	5.7	6.2	6.4	6.6	6.5
Idaho	Boise	7.5	7.2	6.8	6.2	5.8	4.7	2.7	3.1	3.5	5.0	6.6	7.6	5.6
	Pocatello	7.7	7.5	6.8	6.3	6.0	4.8	3.4	3.4	3.8	5.1	6.5	7.7	5.8
Illinois	Cairo	6.5	6.3	6.3	6.0	5.9	5.6	5.2	5.0	4.6	4.4	5.5	6.3	5.6
	Chicago	6.4	6.2	6.1	5.8	5.3	5.2	4.3	4.6	4.7	5.0	6.3	6.6	5.5
	Springfield	6.4	6.4	6.4	6.1	5.9	5.7	4.7	4.9	4.9	4.8	6.1	6.6	5.7
Indiana	Ft. Wayne	7.[illegible]	7.0	6.6	6.4	5.8	5.6	4.7	5.0	5.0	5.4	6.7	7.4	6.1
	Indianapolis	7.[illegible]	6.8	6.6	6.4	6.0	5.6	5.0	5.1	4.8	5.0	6.3	7.1	6.0
Iowa	Des Moines	5.[illegible]	5.5	5.8	5.7	5.6	5.3	4.3	4.4	4.5	4.4	5.3	5.8	5.2
	Dubuque	6.[illegible]	6.0	6.2	5.9	5.9	5.6	4.7	5.0	5.1	5.2	6.3	6.5	5.7
	Sioux City	5.8	5.8	6.1	5.9	5.9	5.3	4.3	4.6	4.5	4.6	5.7	5.9	5.4
Kansas	Concordia	5.[illegible]	5.8	5.9	6.0	6.0	5.0	4.5	4.2	4.0	4.0	4.8	5.3	5.1
	Dodge City	4.[illegible]	4.5	4.7	4.8	4.9	4.3	3.9	3.7	3.5	3.4	3.6	3.9	4.1
	Goodland	5.8	5.9	6.0	6.0	6.2	4.4	4.4	4.4	3.6	3.9	4.8	5.1	5.0
	Wichita	5.0	5.1	5.2	5.3	5.3	4.6	3.9	3.8	3.9	4.0	4.4	5.0	4.6
Kentucky	Louisville	6.8	6.3	6.1	6.0	5.5	5.3	4.8	4.8	4.5	4.6	5.7	6.7	5.6
Lousiana	New Orleans	6.0	6.0	5.8	5.3	5.1	5.4	6.1	5.8	5.2	4.0	4.7	5.9	5.4
	Shreveport	5.8	5.7	5.4	5.1	4.8	4.4	4.5	4.0	4.0	3.7	4.7	5.6	4.8
Maine	Caribou	7.2	6.8	6.8	7.3	7.4	7.5	7.0	6.8	6.6	6.8	8.1	7.3	7.1
	Eastport	6.9	6.5	6.6	6.7	6.9	6.9	6.5	6.3	6.3	6.6	7.6	7.2	6.8
Massachusetts	Boston	6.0	5.6	5.6	5.8	5.8	5.7	5.6	5.3	5.1	5.3	6.0	6.0	5.7
Michigan	Alpena	7.7	6.9	6.4	6.0	5.8	5.5	4.8	5.1	5.8	6.3	7.9	8.0	6.4
	Detroit	7.4	6.9	6.5	6.1	5.7	5.2	4.7	4.8	5.0	5.4	7.1	7.6	6.0
	Grand Rapids	8.2	7.5	6.8	6.3	5.9	5.5	4.7	5.0	5.4	6.0	7.5	8.3	6.4
	Marquette	7.7	7.3	6.9	6.3	6.3	5.9	5.5	5.8	6.3	7.0	8.0	8.0	6.8
	Sault Ste. Marie	7.9	6.8	6.3	5.9	6.0	5.5	5.2	5.6	6.3	7.2	8.4	8.3	6.6
Minnesota	Duluth	6.1	5.5	5.6	5.5	5.8	5.6	4.8	5.0	5.4	5.9	6.8	6.1	5.7
	Minneapolis	6.5	6.2	6.7	6.5	6.4	6.0	4.9	5.1	5.1	5.4	6.9	6.9	6.1
Mississippi	Meridian	6.1	6.0	5.5	5.1	4.9	4.9	5.6	4.9	4.5	3.9	4.6	6.1	5.2
	Vicksburg	5.7	5.8	5.6	5.3	5.1	4.8	5.5	4.9	4.1	3.4	4.5	5.6	5.0
Missouri	Kansas City	5.4	5.4	5.5	5.5	5.3	4.9	4.1	4.1	4.2	4.1	5.0	5.5	4.9
	St. Louis	6.3	5.9	6.0	5.7	5.3	5.0	4.2	4.5	4.1	4.1	5.4	6.4	5.2
	Springfield	5.7	5.3	5.2	5.0	4.6	4.1	3.6	3.9	3.8	4.0	4.8	5.6	4.6

continued on page 8.

Table 1-1. Mean Sky Cover for Selected Cities, Sunrise to Sunset (10 Represents Complete Sky Cover). (Continued from page 7.)

Location		by Month												Annual
		J	F	M	A	M	J	J	A	S	O	N	D	
Montana	Havre	6.3	5.9	5.9	5.7	5.5	5.3	3.6	3.8	4.8	5.4	6.2	6.3	5.4
	Helena	6.8	6.6	6.6	6.5	6.5	6.0	4.0	4.2	5.1	5.7	6.6	6.8	6.0
	Kalispell	7.7	6.7	6.2	5.7	5.6	5.3	3.5	3.9	4.9	5.7	7.6	8.0	5.9
	Miles City	6.8	6.9	7.0	6.8	6.3	5.7	3.9	4.1	5.2	5.3	6.6	6.5	5.9
Nebraska	Lincoln	5.5	5.7	5.8	5.9	5.9	5.2	4.3	4.6	4.4	4.4	5.3	5.6	5.2
	North Platte	6.2	6.2	6.4	6.6	6.6	5.3	4.9	4.9	4.5	4.4	5.8	5.8	5.6
Nevada	Ely	6.2	6.1	6.2	6.0	5.9	4.0	3.9	3.4	3.0	4.1	5.4	6.0	5.0
	Las Vegas	5.2	4.2	4.1	3.8	3.3	1.6	2.8	2.3	1.5	2.3	3.2	4.4	3.2
	Reno	5.3	5.3	4.9	4.6	4.1	3.1	1.8	1.8	2.1	3.4	4.3	5.5	3.9
	Winnemucca	6.4	6.3	6.0	5.8	5.1	4.0	2.3	2.2	2.4	4.0	5.3	6.6	4.7
New Hampshire	Concord	5.6	5.0	4.9	5.2	5.2	4.8	4.7	4.8	5.0	5.3	6.0	5.9	5.2
New Jersey	Atlantic City	6.2	5.6	5.7	5.8	5.6	5.5	5.3	5.2	5.1	4.8	5.3	6.0	5.5
New Mexico	Albuquerque	4.3	4.5	4.6	4.5	4.3	3.4	4.4	4.5	3.2	2.9	3.1	4.0	4.0
	Raton	5.1	5.4	5.2	5.9	6.1	4.5	5.3	5.4	3.9	4.0	4.4	4.2	5.0
	Roswell	4.1	4.0	4.0	3.8	3.9	3.2	3.6	3.6	3.4	3.3	3.3	3.9	3.7
New York	Albany	6.6	6.0	5.9	5.8	5.6	5.3	5.1	5.0	5.0	5.4	6.4	6.7	5.7
	Binghamton	7.5	7.2	7.1	6.9	6.5	6.2	6.1	6.0	6.0	6.3	7.5	7.9	6.8
	Buffalo	8.1	7.5	7.0	6.5	6.0	5.5	5.2	5.4	5.6	6.2	7.7	8.2	6.6
	Canton	6.9	6.3	6.0	6.1	5.7	5.0	4.8	5.0	5.5	6.1	7.5	7.4	6.0
	New York	6.2	5.8	5.9	5.2	5.8	5.7	5.6	5.2	5.0	5.0	5.8	6.0	5.7
	Syracuse	7.9	7.5	7.1	6.6	6.0	5.7	5.4	5.6	5.8	6.3	7.7	8.1	6.6
North Carolina	Asheville	6.0	5.7	5.6	5.4	5.3	5.5	5.9	5.7	5.2	4.4	4.8	5.8	5.5
	Hatteras	6.1	5.9	5.9	5.1	5.3	5.4	5.8	5.6	5.7	5.1	5.1	5.9	5.6
	Raleigh	6.0	5.5	5.4	5.1	5.1	5.4	5.6	5.5	5.2	4.4	4.7	5.8	5.3
	Wilmington	5.7	5.4	5.3	4.6	4.9	5.3	5.6	5.4	5.6	4.4	4.5	5.4	5.2
North Dakota	Bismarck	6.6	6.6	6.9	6.6	6.4	6.2	4.5	4.8	5.4	5.6	6.6	6.6	6.1
	Devils Lake	6.1	6.0	6.1	5.8	5.8	5.6	4.5	4.7	5.3	5.6	6.6	6.3	5.7
	Fargo	6.5	6.3	6.4	6.0	5.7	5.7	4.4	4.6	5.3	5.7	6.7	6.7	5.8
	Williston	6.8	6.9	7.1	6.6	6.4	6.6	4.7	5.0	5.9	6.0	6.9	6.9	6.3
Ohio	Cincinnati	7.4	6.8	6.8	6.5	6.2	5.8	5.1	5.1	4.8	5.0	6.4	7.3	6.1
	Cleveland	7.7	7.4	6.7	6.1	5.6	5.2	4.6	4.8	5.1	5.7	7.4	8.0	6.2
	Columbus	7.1	6.7	6.6	6.2	5.6	5.2	4.8	4.8	4.7	5.0	6.5	7.2	5.9
Oklahoma	Oklahoma City	5.5	5.7	5.4	5.5	5.6	4.7	4.2	4.3	4.1	4.1	4.6	5.2	4.9
Oregon	Baker	6.9	6.7	6.3	6.9	5.7	5.0	2.8	2.8	3.8	4.5	6.0	6.7	5.3
	Medford	8.2	7.6	7.2	6.6	5.8	4.9	2.1	2.2	3.3	5.6	7.4	8.6	5.8
	Portland	7.9	7.9	8.0	7.4	7.1	6.9	4.6	5.4	5.1	7.0	8.2	8.6	7.0
	Roseburg	8.9	8.5	7.9	7.2	6.4	6.4	2.8	3.7	4.9	6.9	8.4	8.9	6.7
Pennsylvania	Harrisburg	6.6	6.1	6.1	6.2	6.0	5.9	5.6	5.6	5.1	5.2	6.2	6.5	5.9
	Philadelphia	6.1	5.8	5.8	5.8	5.8	5.8	5.7	5.6	5.2	4.9	5.6	6.0	5.7
	Pittsburgh	7.4	7.0	6.7	6.3	5.7	5.5	5.3	5.2	5.0	5.4	6.7	7.5	6.1

Rhode Island	Block Island	6.0	5.5	5.5	5.7	5.7	5.3	5.2	5.0	4.8	4.6	5.7	5.9	5.4
	Providence	5.9	5.4	5.4	5.7	5.6	5.4	5.4	5.1	5.0	4.7	5.5	5.7	5.4
South Carolina	Charleston	5.4	5.3	5.1	4.5	4.6	5.3	5.7	5.5	5.1	4.3	4.3	5.3	5.0
	Columbia	6.0	5.7	5.8	5.1	5.2	5.1	5.8	5.1	5.6	4.6	4.8	5.7	5.4
South Dakota	Huron	6.6	6.4	7.2	6.6	6.2	5.7	4.6	4.9	5.0	5.1	6.6	6.7	6.0
	Rapid City	6.4	6.4	6.6	6.6	6.5	5.7	4.2	4.3	4.6	4.9	6.2	6.2	5.7
Tennessee	Knoxville	5.9	5.8	5.5	5.2	4.9	5.0	5.1	5.0	4.1	3.7	4.7	5.7	5.0
	Memphis	7.2	6.5	6.5	6.0	6.0	5.2	5.3	4.7	4.5	4.1	5.2	6.2	5.6
	Nashville	6.8	6.5	6.1	5.8	5.6	5.2	5.2	5.0	4.7	4.3	5.5	6.5	5.6
Texas	Abilene	5.6	5.6	5.4	4.9	5.2	4.2	4.2	4.2	4.4	3.8	4.3	5.1	4.7
	Amarillo	4.2	4.5	4.2	4.3	4.4	3.8	4.0	3.8	3.7	3.7	3.6	4.2	4.0
	Austin	6.2	6.1	5.7	5.6	5.5	4.7	4.5	4.0	4.5	4.2	5.1	5.8	5.2
	Brownsville	6.5	6.2	6.1	5.9	5.4	4.7	4.5	4.3	5.0	4.4	5.7	6.4	5.4
	Corpus Christi	6.5	6.7	6.3	6.2	5.8	4.6	4.5	4.2	4.7	3.9	5.2	6.4	5.4
	Del Rio	5.2	5.3	5.3	5.4	5.7	4.9	4.2	3.5	4.4	4.6	5.1	5.2	4.9
	El Paso	3.7	3.6	3.4	2.9	2.6	2.4	4.0	3.8	3.1	2.7	2.8	3.5	3.2
	Ft. Worth	5.9	5.6	5.2	5.2	5.4	4.2	4.0	3.8	3.7	3.9	4.4	5.3	4.7
	Galveston	5.9	5.9	5.7	5.4	4.8	4.1	4.7	4.5	4.3	3.6	4.7	5.9	5.0
	Houston	6.1	6.1	5.9	5.7	5.4	4.8	5.2	5.0	4.7	4.2	5.2	6.2	5.4
	Midland	5.4	5.3	5.1	4.9	4.7	3.8	4.6	3.9	3.5	3.6	3.5	4.3	4.4
	San Antonio	6.4	6.5	6.3	6.4	6.2	5.4	5.0	4.6	4.9	4.6	5.4	5.7	5.6
Utah	Modena	5.2	5.5	5.0	5.1	4.4	2.8	3.7	3.5	2.8	3.6	4.1	5.0	4.2
	Salt Lake City	6.9	7.0	6.5	6.1	5.4	4.2	3.5	3.4	3.4	4.3	5.6	6.9	5.3
Vermont	Burlington	7.2	6.9	6.6	6.7	6.7	6.1	5.8	5.7	6.0	6.5	7.9	7.8	6.7
Virginia	Norfolk	5.6	5.2	5.1	4.8	4.8	4.9	5.0	5.0	4.6	4.4	4.6	5.4	5.0
	Richmond	5.9	5.2	5.3	5.1	5.0	5.2	5.2	5.2	4.9	4.3	4.8	5.5	5.1
	Roanoke	6.7	6.3	6.4	6.3	6.3	5.8	6.0	5.9	5.5	5.2	5.3	6.2	6.0
Washington	North Head	7.8	7.3	7.1	6.9	6.9	6.8	6.4	6.5	6.2	6.8	7.6	7.8	7.0
	Seattle	8.0	7.7	7.4	6.9	6.4	6.4	4.9	5.3	5.6	7.2	8.0	8.1	6.8
	Spokane	8.1	7.4	6.8	6.4	6.2	5.8	3.5	3.6	4.8	6.0	7.7	8.2	6.2
	Tatoosh Island	7.9	7.3	7.3	7.2	7.1	7.1	6.9	7.0	6.5	6.9	8.0	7.9	7.3
	Walla Walla	8.1	7.4	6.3	5.4	5.0	4.5	2.4	2.7	4.0	5.2	7.4	8.4	5.6
	Yakima	7.7	7.1	6.5	6.1	5.6	5.0	2.6	2.8	3.9	5.4	7.1	7.8	5.6
West Virginia	Elkins	7.9	7.6	7.4	7.0	6.8	6.7	6.5	6.4	6.1	6.1	6.9	7.6	6.9
	Parkersburg	7.3	6.8	6.4	6.0	5.5	5.2	5.0	5.1	4.8	5.1	6.6	7.3	5.9
Wisconsin	Green Bay	5.8	6.5	6.4	6.4	6.4	6.2	5.6	5.7	5.9	6.3	7.2	7.1	6.4
	Milwaukee	5.5	6.3	6.3	6.0	5.7	5.5	4.5	4.8	5.1	5.5	6.5	6.7	5.8
Wyoming	Cheyenne	4.8	5.6	5.7	6.2	6.3	5.3	4.8	5.0	4.5	4.4	5.0	5.1	5.2
	Lander	4.7	4.8	5.3	5.6	5.5	4.5	4.1	4.0	4.0	4.3	4.9	4.7	4.7
	Sheridan	6.5	6.7	6.9	6.6	6.6	5.8	3.9	4.1	4.7	5.1	6.4	6.6	5.8
	Yellowstone	5.9	6.4	6.4	6.1	6.3	5.4	4.3	4.3	4.9	5.4	6.4	6.8	5.8
Puerto Rico	San Juan	[illegible]	4.5	4.5	5.0	5.8	5.8	5.5	5.1	5.8	5.4	5.2	5.1	5.2
Virgin Islands	St. Croix	[illegible]	5.8	5.7	5.9	7.2	6.9	7.0	6.1	7.0	6.5	6.0	5.8	6.3

Source: Climatic Atlas of the United States

The Heat Box

A heat box is similar in construction to a flat-plate solar collector. Both are covered with one or more flat pieces of glass or clear plastic that serves to trap heat inside an insulated box. In the heat box, however, it is advantageous to have a durable metal liner, the inside of which is coated with a heat-resistant flat-black paint. The food is placed inside the box in pots or roasters with tight-fitting covers. These containers should be black for maximum energy absorption.

The heat box will operate in either direct or diffuse sunlight and will continue to increase in temperature until equilibrium, or a heat balance, is obtained. As the temperature in the box increases, more heat will be lost through the glass. This heat loss may be reduced by adding more covers. Experiments have shown that two panes of glass spaced approximately ½-inch apart provide the best efficiency for the least money. The heat box in Fig. 1-4 is designed to be used in a horizontal position. As you may have already learned from solar collector theory, maximum heat is obtained from south-facing collectors when they are inclined perpendicular to the sun's rays. It is possible to tilt the heat box to line up with the sun; but this is often impractical. The food can slide and spill inside the cooking chamber. There are better ways to increase the usefulness of the box, and these will be discussed in the next section.

Under the best conditions, a heat box cannot be expected to provide temperatures much above 200° F. Therefore, its use is limited to areas where the sun travels in a high arc and can shine directly on the food containers without extensive shadows from the sides of the box. With this restriction, it is best used as a container to keep food warm once it has been cooked on a concentrator.

Reflector

By adding reflectors to the heat box and inclining the glass covers, you can overcome most of the disadvantages of the heat box. This device is commonly called the *Telkes oven,* named after the noted physicist Dr. Maria Telkes. These are not curved reflectors that focus energy as the concentrators described earlier. They are flat and serve to augment the solar energy striking the cover glass of the heat box. Generally, they are made from flat pieces of highly specular aluminum, such as Alzak (made by Alcoa), or silvered glass mirrors. Four or more reflectors are attached to the top of the heat box with hinges that allow them to be removed, folded for storage, or adjusted for maximum heat gain. Angling them at 60° to the plane

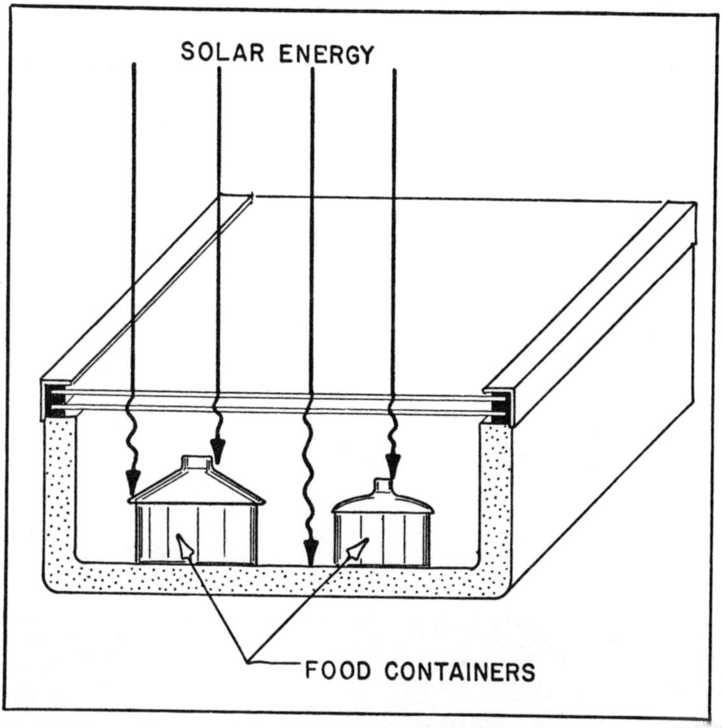

Fig. 1-4. This typical heat box with dual glazing and metal liner is designed to be used in a horizontal position.

of the glass cover provides good results (Fig. 1-5). When the sky is clear (a high percentage of direct radiation), the amount of solar energy striking the absorber can be increased one and a half to two times by using four reflectors, each of approximately the same area as the cover glass. This will add enough heat to reach oven temperatures of 350° to 400° F.

Because the arc of the sun is not directly overhead except at the equator, the efficiency of the solar oven can be improved if you incline the glass cover at an angle that will intercept a maximum amount of solar energy. For south-facing collectors, this occurs when the cover glass is at *right angles* to the sun. For example, if the sun's elevation of altitude angle is 60° above the horizon, tilting the cover 30° would provide a right angle and maximum energy absorption. One method of accomplishing this inclination while keeping the cooking pot level is suggested by the triangular oven design and cradle support shown in Fig. 1-6. The cradle support permits minor adjustments to maximize efficiency as the sun's elevation changes

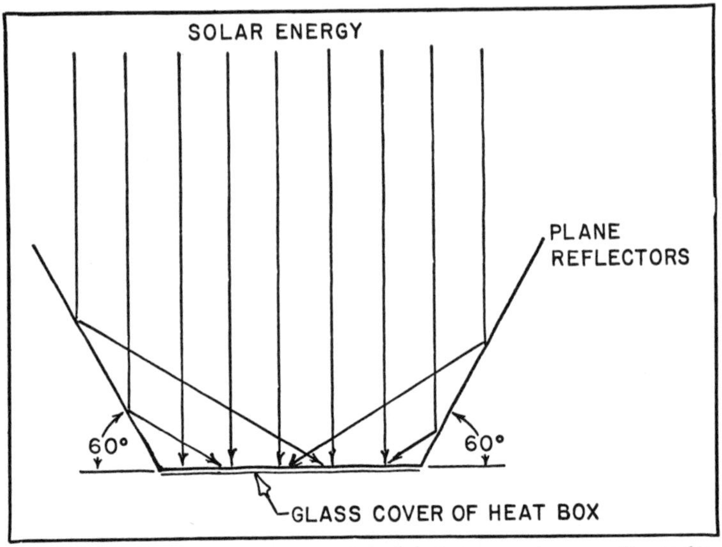

Fig. 1-5. Flat reflectors inclined at 60° to the glass cover increase the performance of the heat box.

Fig. 1-6. Solar oven with cradle support (courtesy, Dr. Maria Telkes).

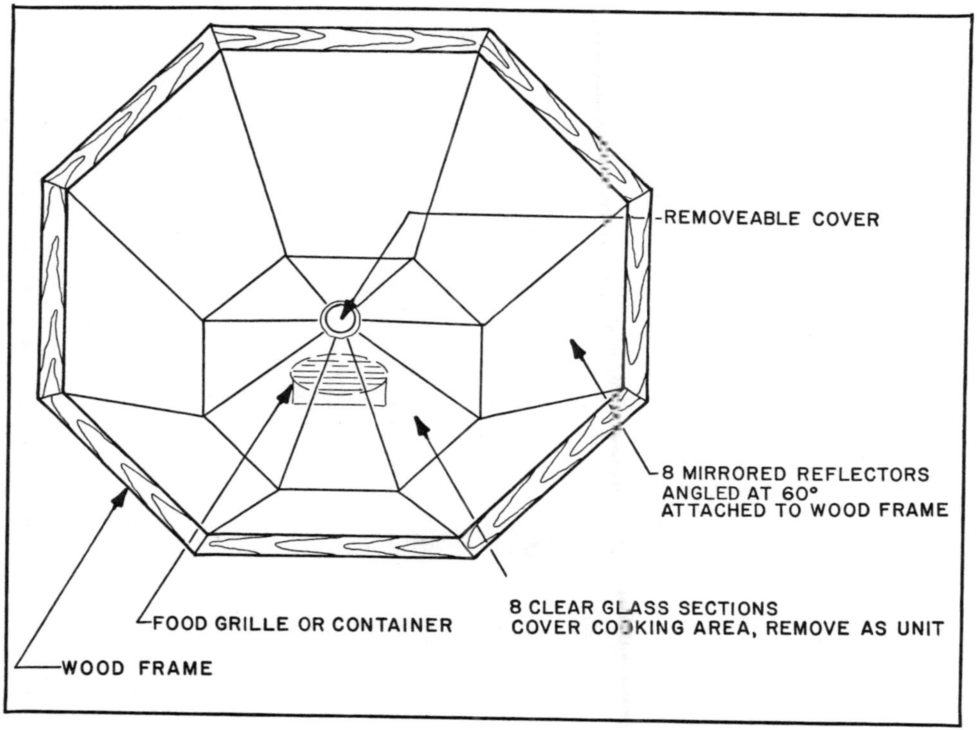

Fig. 1-7. An octagonal reflector oven is a practical and compact solution.

throughout the day. With this feature, adjustment every one-half hour is suggested.

If you plan to build an oven of this type, I would suggest that you incline the glass at 10° *less* than your latitude for summer cooking and at 10° greater than your latitude for winter cooking, and at latitude for year-round use (see Figs. 1-10 and 1-11). You can obtain the position of the sun for a specific latitude, hour, and day of the year from almanacs or other sources. Most researchers agree that for latitudes above 40° north, where the angle of solar radiation (altitude) is less than 25° during half of the year, the solar oven is not very practical.

There is another approach to the reflector oven. It was conceived and patented in 1876 by W. Adams, a cook for the British Army stationed in India. Eight reflectors form an octagon that surrounds the cooking pot rather than being placed above it. The reflectors are fixed, but the entire unit can be tilted to intercept more sunshine (Fig. 1-7). This is a unit you can design yourself using the basic principles discussed, or you can purchase it in kit form.

DESIGN CONSIDERATIONS

Following are some basic principles that professionals have used in solar cooker designs. Knowing these principles will help you to understand how these devices work and may even motivate you to design your own cooker.

Concentrators

Traditional designs use a parabolic reflector to focus solar energy on the food or pot containing it. The pot is placed at the focal point of the parabola. It may be helpful to review the characteristics and measurement of the parabola. If you were to cut through a right circular cone at any height and parallel to the base a *circle* would be formed (Figs. 1-8 and 1-9). If the cut is made parallel to either side of the cone, the result is a *parabola*. Mathematically the parabola is defined by the following equation:

$$y^2 = 4\,Fx$$

where:

F is the focal point of the parabola.

Solar energy reflected from the parabola is concentrated at the focal point F, where the cooking pot would be located. In the case of a trough-type or line-focusing cooker, this would be the location of the spit or cooking tube. The choice of the focal *length* depends upon the reflector size. A larger reflector will increase the capacity of the cooker, but it is generally not practical to exceed a 4-foot diameter, because larger units are expensive and difficult to handle. Experimenters have tried various focal lengths and found 20 inches to be a good compromise for a 4-foot diameter reflector. With long focal lengths (over 3 feet), it becomes more and more difficult to keep the cooking pot stable and level as the reflector is adjusted to track the sun. Also, as you increase the distance from the reflector, more heat is lost due to convection and radiation.

Some designers prefer to use spherical reflectors. They are generally easier to fabricate and may be lower in cost than the parabolic type. Umbrella (folding) reflectors are made from pie-shaped segments and are close to the spherical form. Parabolic reflectors have a sharp focal point that creates high temperatures over a small area. Advocates of the spherical type advise that this hot spot can burn meat in one area and leave other portions raw. The spherical reflector, on the other hand, provides a diffuse focus and spreads less intense heat over a larger area. This results in more uniform cooking.

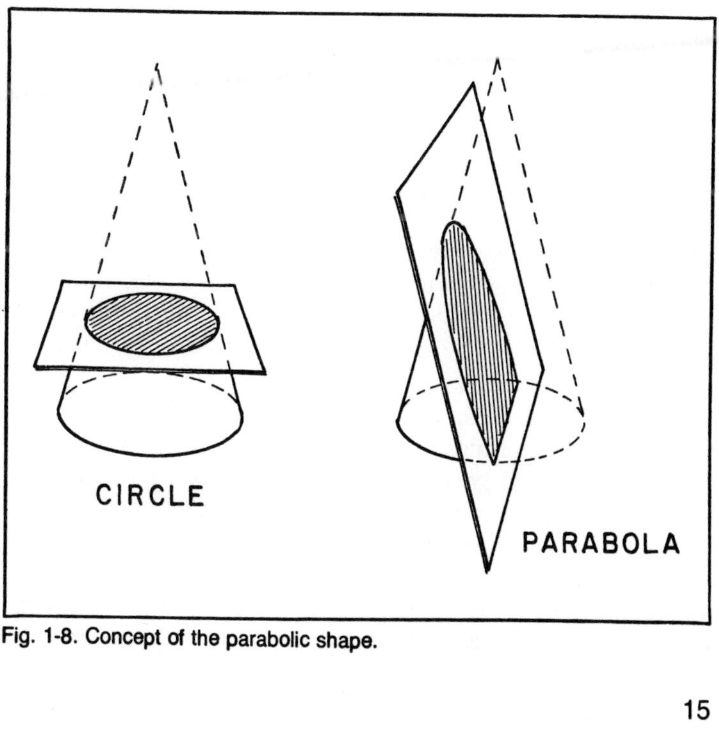

Fig. 1-8. Concept of the parabolic shape.

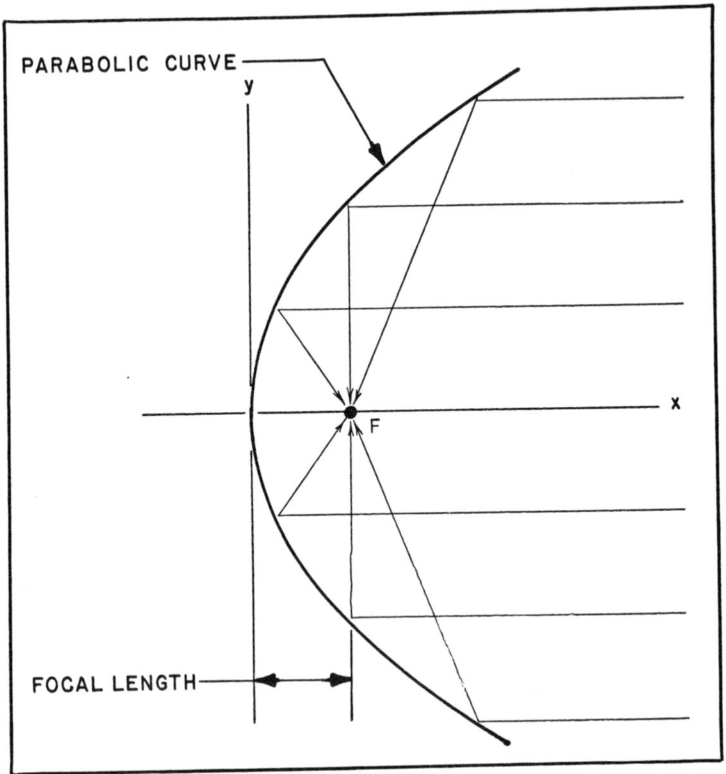

Fig. 1-9. Characteristics of a parabolic curve showing focal length and focal point F.

The output of concentrator cookers will vary between 200 and 400 watts per square meter (10.8 square feet) of reflector surface on a bright, clear day depending upon the quality of the reflective surface and optical accuracy. Most designs target for an output of 500 watts, about 50 percent of what an average gas or electric stove top burner will deliver. With this output, you should be able to boil a quart of water in 15 minutes. This boiling test provides a good method for comparing commercial units.

Most reflector materials will lose a portion of their original specularity (reflection ability) due to weathering and the necessary cleaning of food and grease splatter. Aluminized Mylar plastic, cemented to a rigid plastic shell is commonly used due to price, but the material cannot be expected to last more than two years with regular use. Spun aluminum has been used for many years as a reflector material in the lighting and electronics industries. These metal reflectors offer better durability, but they require protective coatings to retain good specularity. The Umbroiler cooker uses an aluminized Mylar plastic/rayon-laminated cloth reflector which can be folded for storage. However, it is difficult to focus sharply because it flutters when wind is present. Thin strips of mirrored glass cemented along the parabola are sometimes suggested for trough-type concentrators. I would not recommend them for this type of unit, however, as food spills and grease collect at the edges and become difficult, if not impossible, to clean.

Reflector Ovens

Much of the published data on solar radiation relate to the amount of energy falling on a horizontal surface. As noted earlier, the sun is seldom directly overhead, and for maximum efficiency the collector surface or glass cover should be perpendicular to the sun. You can get a feel for the importance of this factor from Fig. 1-10.

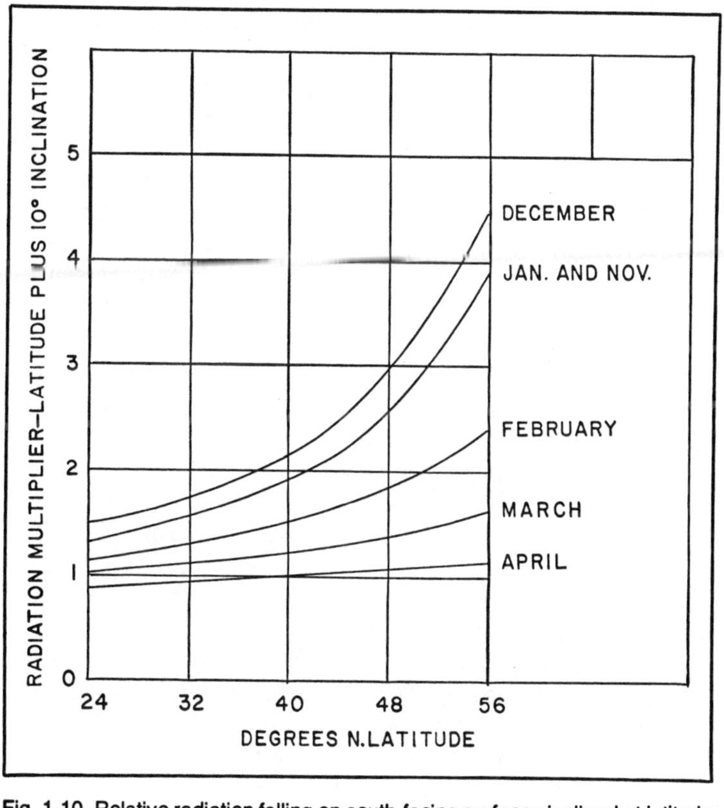

Fig. 1-10. Relative radiation falling on south-facing surfaces inclined at latitude plus 10°.

For example, if your location is 40° north latitude, during the month of January a collector inclined at 50° (latitude plus 10°) would intercept twice as much solar radiation as one mounted horizontally or flat. (The radiation multiplier in this case is 2). This effect is most important in northern latitudes during winter months. Another reason to incline the glass is so that you keep reflectance to a minimum. The amount of energy transmitted through window glass is directly related to the angle at which the sunlight strikes the glass. This is called the *angle of incidence* and is measured as a deviation from a perpendicular line extending from the glass surface. Transmittance falls off quite rapidly at incident angles greater than 60° (Fig. 1-11). For these reasons, angling the glass will improve the performance of the reflector oven over a horizontal design.

When direct solar energy is available, flat reflectors inclined at approximately 60° to the glass cover of the triangular oven will increase the amount of solar energy falling on the glass (see Fig. 1-5). Experiments have shown that with four square reflectors, each having the same area as the cover glass, the amount of energy inside the oven can be increased by a factor of 1.6 to 1.8 depending upon the percentage of direct radiation and number of cover panes used. Oven temperatures up to 300° F are attainable with two glass covers. Each layer of glass improves the insulating value of the oven by reducing heat losses, but remember that the glass actually reflects some of the solar energy striking it and absorbs some, thereby reducing the amount transmitted to the cooking pot inside the oven.

Use Table 1-2 for developing the best design. These data were collected by Dr. Telkes, using an oven with a glass window area of 1 square foot and four Alzak reflectors of equal size. Temperatures were recorded on a blackened aluminum plate placed inside the oven. Four-inch thick fiberglass insulation was used throughout. A ½-inch air space separated each layer of glass. Similar tests were made using silvered-glass mirrors, in place of the Alzak reflectors. The mirrors provided an average temperature increase of 10° F but are fragile and add weight to the unit. If you do not foresee this as a problem, then mirrored glass should be considered. You may conclude, as I did, that one cover is not enough. Two appear to provide the best value when cost of construction is considered.

In cooking, most of the energy required is used to bring the food up to cooking temperature. Very little is needed to sustain that level in a covered pot. For example, to boil water using an electric

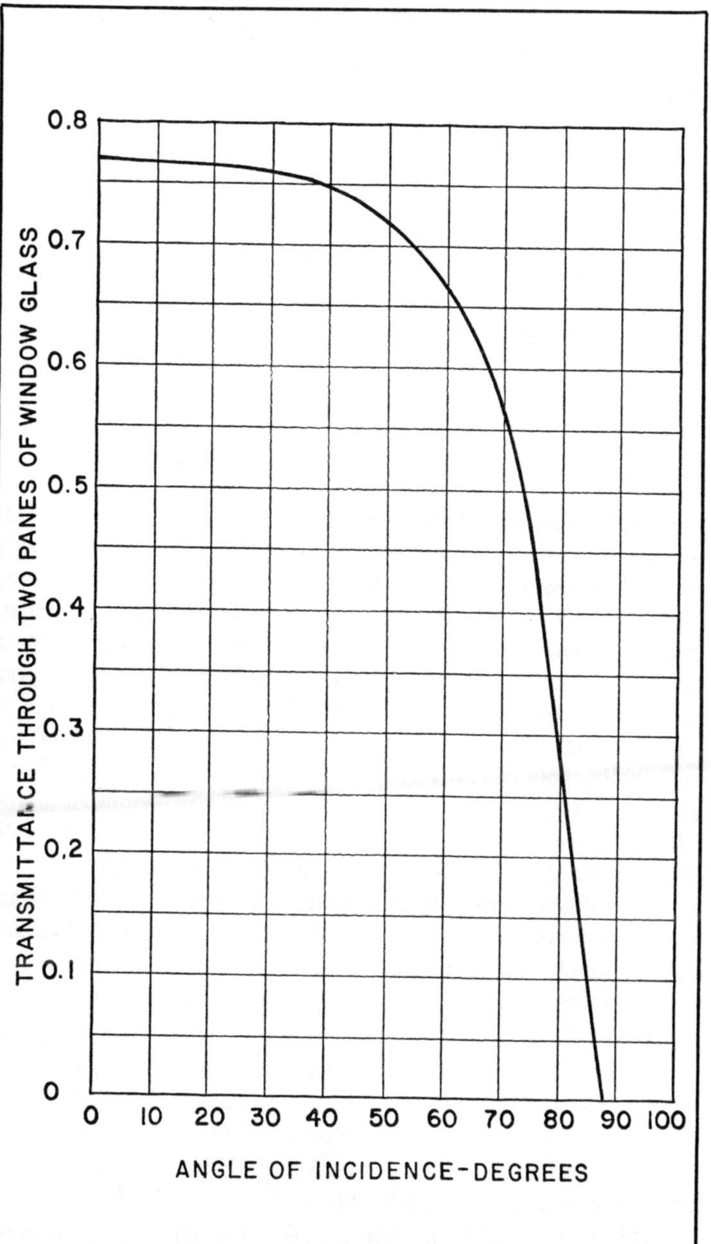

Fig. 1-11. Energy loss as a function of nonperpendicular glazing surfaces. If the cover glass is perpendicular to sun angle (0° incidence), 77 percent of the solar energy that strikes the glass will be transmitted into the oven.

Table 1-2. Effect on Cooking Temperature by Adding Reflectors and Multiple Glass Covers.

Number of Reflectors	Number of Glass Covers	Test #1		Test #2	
		Insolation Btu/sq ft-hr	Average Plate Temp. (°F)	Insolation Btu/sq ft-hr	Average Plate Temp. (°F)
0	1	234	244	242	242
	2	264	303	289	288
	3	280	335	270	320
4	1	234	348	264	360
	2	286	380	277	420
	3	276	434	272	430

stovetop burner, you would start with the high setting. Once boiling begins, the burner is turned down to low or simmer, and the boil will continue as long as the pot is covered and water does not evaporate. The same holds true for foods, most of which require a similar amount of heat. It helps, then, to have an oven that can be preheated and will then store heat, since food absorbs heat quite rapidly upon entering the oven. Solar ovens with a window approximately 2 square feet in size can store 80 or 90 Btu in the metal interior of the oven walls. This heat storage may be supplemented by a cooking platform filled with phase-change materials. Six pounds of anhydrous alkaline sulfate will hold up to 600 Btu of heat. This will increase the storage capacity of the oven up to 690 Btu.

Dr. Telkes performed cooking tests on several foods using an oven of this type and reported the results to the United Nations Conference on New Sources of Energy in 1961. Here are the results of those tests:

Rice. One pound of rice and 1.2 pounds of water in a flat pan cooked in 45 minutes without stirring or other attention. The water was absorbed completely and the rice was perfectly cooked.

Lentils. One pound of lentils required 4 pounds of water. The mixture stood at room temperature for 12 hours to soften the lentils. Cooking time was 2 hours.

Dry peas and black beans. One pound required 4 pounds of water. It was softened for about 12 hours. The peas and beans required 3 to 4 hours cooking time, until they were sufficiently tender.

Roasts. The oval roaster held up to 8 pounds of meat (beef, veal, pork, etc.). Roast beef (8 pounds) required 3 hours. Two chickens (4 pounds) were completely roasted in one hour. On clear days the roasting time was approximately the same as in conventional ovens.

Stews. Made of meat and vegetables, stews required about 2 hours cooking until the meat was sufficiently tender.

Bread, Rolls, and Cakes. Two loaves of bread (2 pounds) baked in 45 minutes; rolls required about 30 minutes; cake (3 pounds) one hour. The baked food was uniform in texture, and the results were comparable to food baked in conventional ovens.

Fruit Preserves. Two pounds of fruit were mixed with one pound of sugar and 0.5 pound of water. Cooking time was 3 hours. The preserves were put into containers and sterilized in the oven, in the usual way. These tests indicate that the solar oven can be used for the preparation of preserved and "home-canned" foods.

Under bright, clear skies, the solar oven cooked at about the same rate as a conventional gas or electric unit. Slightly longer times are required when partial clouds or haze are present.

HOME FOOD DRYING

Almost every kind of fresh food will, in time, deteriorate. Spoilage is increased in moist, warm weather. Certain foods will keep for months in cool, dry places, but eventually they, too, will fall victim to the rotting process. Acid-based foods are slower to deteriorate than alkaline based food. Fruits are excellent candidates for drying and can be kept for several months once their moisture has been removed. *Removing moisture* is the key to preserving fresh foods, for without moisture, bacteria cannot grow at a harmful rate.

Fruits are usually dried in the sun until their moisture content is down to 5 to 10 percent. Some fruits can be preserved for a year or more if kept in this dry state. The natural sugar in fruits helps to preserve them. Many vegetables are also good candidates for drying. With dried fresh food you can eliminate from your diet those chemicals that are found in most commercially processed canned fruits and vegetables.

Drying Methods

Many of the basic principles involved in solar stove design also apply to drying. You can, for example, use *natural air drying,* placing the food on open racks that face south and are inclined to capture more of the sun's energy (see Fig. 1-10). The drying may be accelerated with the use of flat reflectors on the sides of the rack (see Fig. 1-5). In this case, there is no cover glass and the food is placed skin down on a porous surface. The moisture vapor that is produced is removed by air flowing over and around the food. This method leaves a lot to be desired, however, because the food must be covered during nondaylight hours or rainy weather, and it must be protected from dirt, dust, and insects.

The solar *cabinet dryer* provides a solution to many of these problems. In principle, the dryer is like a heat box with an air intake and exhaust port for moisture removal. Food is placed on shallow trays inside an enclosed cabinet (Fig. 1-12). Heat passes through the glass or plastic window (called glazing) and is trapped. This causes moisture to evaporate from the food. If there were no means to remove this moisture, the food would cook, not dehydrate. Ports or vents provided at the base and top of the cabinet allow the hot, moisture-laden air to escape, and cooler, dryer air enters to replace it. This air movement takes place naturally, no fan or blower is needed. The air flow can be increased by making the upper opening larger—the principle used in my design. Ventilation openings are covered with screening to keep insects out.

HOME-BUILT CABINET DRYER

The design illustrated in Fig. 1-12 is for a relatively small two-tray dryer with a base dimension of approximately 19 inches by 24 inches and an 18-inch height. It is an inexpensive unit that can easily be built over a weekend.

The enclosure is constructed of ½-inch thick exterior-grade or marine plywood. As no dimension exceeds 2 feet, it is quite possible that you can purchase remnant pieces rather than a full 4-by-8-foot sheet. Be sure the material is marked "exterior." The letters stamped on plywood refer to the grades of veneer on the faces. A is the best, and D, the lowest. Exterior-grade wood generally uses grade C for the inner plies. Assembly is made using waterproof glue and plated screws or nails for additional support as needed.

Glazing may be either double-strength window glass (about 0.12-inch thick) or a premium grade of fiberglass with good solar transmission (0.040-inch thick). If fiberglass is selected, it will deflect more sunlight than will glass. It will need two vertical supports over the 2-foot span at the screened opening at the rear of the enclosure. I would suggest one support in the center for glass.

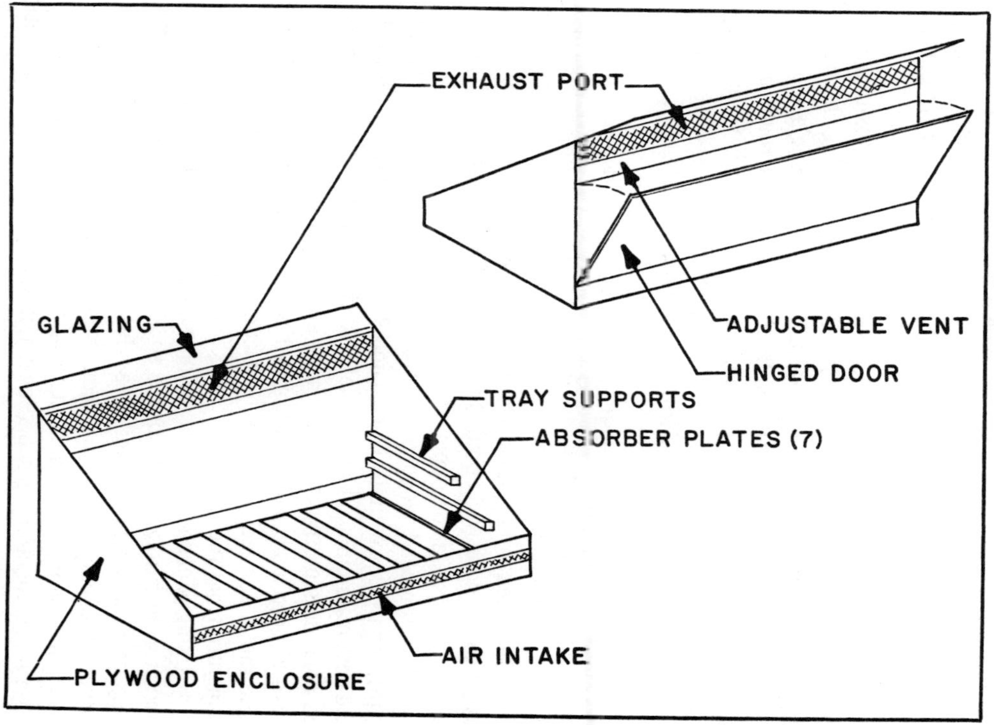

Fig. 1-12. Home-built cabinet dryer Food trays are not shown.

Glazing is attached to the enclosure using a bead of silicone sealant (G.E. Silglaze or equal) applied with a caulking gun. For best adhesion, unfinished wood should be sealed with paint, and all surfaces should be clean and dry. In the design shown, the glazing overhangs the enclosure at the rear to provide some protection from weather, but this is optional. If you use glass, be sure to protect this edge for safety reasons.

Nylon mesh window screen is suggested for the two drying trays and ventilation openings. This material is noncorrosive and relatively easy to clean. A screen shop can quickly fabricate the trays from aluminum tube-type frame material with a spline. You can make them yourself from a wooden frame by stretching and stapling the nylon mesh around it. Use silicone to attach the screening over the ventilation openings. In the rear, cement the top of the screen directly to the glazing.

It is important to use weather-resistant and nontoxic materials, but there is considerable latitude in constructing the basic design if you have materials on hand. The slope of the glazing was set at 30°, which, represents the best angle for year-round use at 30° latitudes. You may, of course, decide to vary this to fit your specific location.

To build the dryer you will need the materials listed in Table 1-3.

A cross section of the unit (through the center of the enclosure looking at the side) and measurements are shown in Fig. 1-13. The bottom and all panels are applied to the outside edges of the ends. Be sure to leave enough space between the door and upper back panel so the door won't bind. The detail shows a method of providing adjustment of the upper vent opening.

Assemble all wooden parts of the enclosure and spray all interior surfaces with flat black paint. Also spray the top surfaces of the seven aluminum absorber plates at the same time. This should be done outside the cabinet. These surfaces are painted black to absorb more heat inside the cabinet when the dryer is operating. The metal surfaces hold the heat and radiate it upward to drive moisture from the food.

Using silicone, cement nylon screen inside the front vent opening. Position the enclosure so this edge rests horizontally. Apply a bead of silicone, and press the screen in place. Cement the lower edge of the rear vent screen to the inside surface of the upper rear panel, leaving sufficient material to be attached to the glazing after it is cemented in place. (See detail A, Fig. 1-13). Also attach at this time any glazing supports that you feel will be needed. Make certain that the fasteners do not interfere with the vent adjuster.

Cement the glazing in place by applying a bead of silicone along three sides of the enclosure. Be careful not to smear excess silicone on the glazing because it is difficult to remove.

Wait 24 hours for the silicone on the glazing to fully harden, then attach the upper edge of the rear vent screen to the glazing with silicone. It may help if you (gently) invert the enclosure and rest it on the glazing during this final operation.

I suggest you paint all exposed (exterior) wooden surfaces with high-gloss, clear polyurethane to protect the unit against the weather. Any quality exterior grade paint may be substituted.

As mentioned, the two drying trays may be either purchased as window screen assemblies or you can make them. The lower tray is 16 by 22½ inches. This provides enough clearance for the screens

Table 1-3. Typical Bill of Materials (Trays Excluded).

Quantity	Size	Description
2	2 5/16×18×12¾×½"	sides
1	19×24×½"	bottom
1	1 1/16×24×½"	upper front panel
1	1 1/16×24×½"	lower rear panel
1	¾×24×½"	lower front panel
1	5 9/16×24×½"	door
1	2 7/16×24×½"	upper rear panel
1	2 7/16×24×½"	vent adjustment
2	16×¾×¾"	tray supports
2	14×¾×¾"	tray supports
as required		nylon screen
2	22½×½×½"	absorber spacers
7	16×2½"	aluminum absorber plates
1 pair		door hinges
1 each		magnetic door catch
1 sheet	23×24"	glass or plastic glazing
1 spray can		flat black, nontoxic, heat-resistant paint (Rust-Oleum 7776 or equal)
1 can		clear polyurethane paint (exterior grade)
1 tube		silicone sealant (G.E. Silglaze or equal)
as required		waterproof glue, galvanized nails, plated screws, washers

to slide over the tray supports. You may wish to remeasure the enclosure before you order or build the trays.

Place the aluminum absorbers across the supports (absorber spacers) with space in between for air circulation. Block both vents and allow the assembly to "bake" in full sun for a day. Open the vents and let it sit for another day in full sun. This process will thoroughly cure the paint and silicone before the dryer is put into use.

Remember to always wash the food trays before and after you use them. A thin layer of cooking oil will help prevent messy foods from sticking to the nylon screen.

SOLAR DRYER USE

Always situate the dryer, preferably in an unshaded location, so that the window faces south. For indoor food drying, use a south-facing window. Since the glass cover prevents much of the long-wave heat energy from leaving the dryer container, there is a buildup of heat inside the box. Eventually, without some ventilation a "greenhouse effect" from the condensing moisture given off by the drying food would cook the food instead of drying it. By adjusting the vents in the dryer, you can achieve the best drying temperature. Temperatures of 120° to 140° F (50° C) are required for vegetables and such fruits as peaches, apples, cherries, and bananas. In very hot, dry climates, direct sunlight should be avoided on the first day of drying since the food tends to dry too quickly on the outside, sealing the moisture within. This slows the drying time and can produce rot. On warm days you may want to leave the vents completely open. On cooler days a partially closed vent will be best. Open racks must, of course, be covered at night and during rainy weather (when a source of artificial heat may also be needed).

To be sure the optimal temperature is maintained, place a thermometer inside the dryer where direct sunlight will not strike it. Check it after the first half hour and then occasionally during the drying period. To test for dryness, gently squeeze a handful of the food. It should be slightly springy but not damp to the touch. Generally, most fruits and vegetables will take from one to three days to dry out. After it has been sufficiently dried, place the food immediately in *airtight* containers before it has a chance to absorb moisture from the air. Depending on the length of storage, both plastic bags and glass jars with tight-fitting lids will work fine.

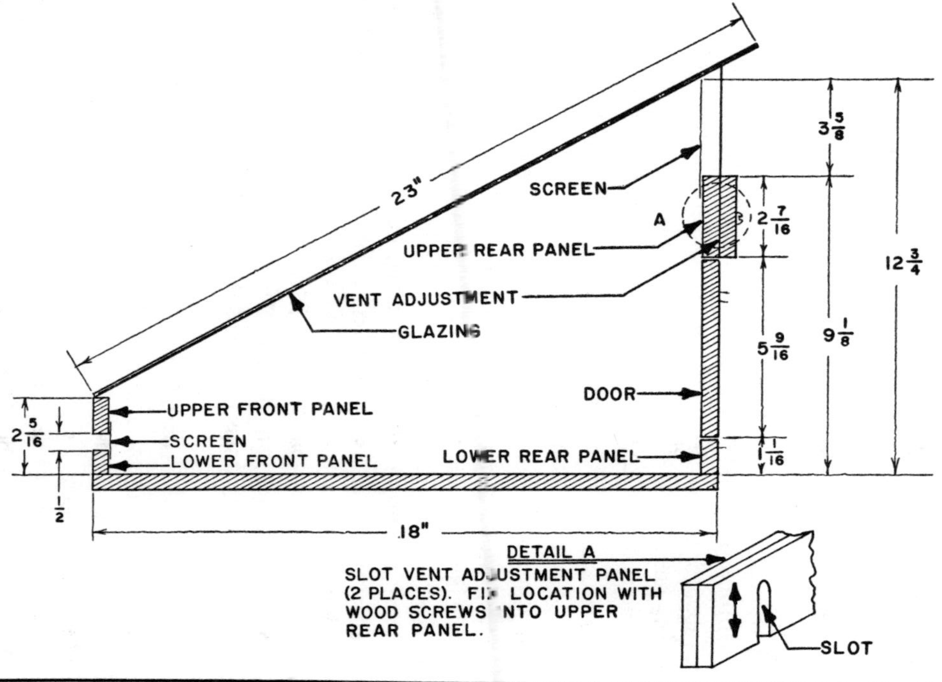

Fig. 1-13. Cross section and construction details of a home food dryer.

Preparing Fruits and Vegetables for Drying

Certain general rules should be observed in preparing your fruits and vegetables for the drying process:

1. Use only completely ripened stock cut into uniform pieces or slices. Whole fruits can take up to a week to dry because of their higher moisture content.
2. If possible, use only fruits and vegetables freshly picked that day.
3. Cut off any bruised or bad spots.
4. Depending on which fruits or vegetables you choose, clean them well, remove the pits, and peel or slice them.
5. Most vegetables require some steaming first to stabilize their color and improve their flavor. Vegetables that have been pre-steamed will take less time to dry.

Part of the fun of solar drying is experimenting with different foods. The following directions for some common foods prepared in the dryer will help you get started.

Onions. Peel the onions. Slice them into uniform horizontal slices about 3/16-inch thick. Place the onions in a single layer on the rack. Dry until brittle. To use in recipes, mix 2 tablespoons of the dried onions with ½ cup of water to get ½ cup of onions.

Garlic. Slice them or use sectioned. Process to a "bone-dry state. Add an equal part of water to reconstitute onions.

Mushrooms. Use only very fresh, firm mushrooms. Wash or wipe them with a damp towel. If they are large, slice them first. Steaming improves the flavor. After drying to a brittle state, they should be "conditioned" for a day or two by placing them in a clean, *dry* area that is protected from dust and insects. This is recommended for mushrooms and other vegetables that may not, because of non-uniform sizes, have been equally dried out. By conditioning them in this manner, the stock is equalized. To reconstitute, soak the dried mushrooms in cold water using one cup of water to every ounce of dried mushrooms (3 ounces dried equals 1 pound of fresh mushrooms).

Zucchini and Cucumber. Scrub the outer surfaces of the vegetables. Slice them, unpeeled, into ¼-inch slices. Place the slices on a rack in a single layer and dry until crisp. Use the slices as is, like potato chips.

Banana. Always use very ripe bananas, peeled, for drying. Cut the bananas into long strips or sections. To use the dried fruit in recipes, soak it in water at a ratio of two cups water to one cup of dried bananas. The dried banana chips can be used for snack food.

Pineapple. Peel and core very ripe pineapple. Use it either sliced or in sections. After it is quite dry, store it in a dry, sterile container.

Fruit Recipes

The following recipes for fruit "leather" and dried fruit cookies are rather basic and may be varied by using different fruits or combinations of fruit.

Fruit Leather. This is sometimes called fruit taffy or fruit roll. Use any fruit you wish. Cut it into chunks. Remove pits or seeds from the fruit. Seeds in berries or grapes can be left. Peeling is optional. Place the fruit in food blender. To yellow or light-colored fruits, add 1 tablespoon lemon juice. Add honey for additional sweetness and purée in blender.

Line a cookie sheet with plastic wrap or spray with a no-stick coating. A dark or black pan works best for this. Be sure the cookie sheet has a raised edge to prevent spillage. Pour the purée on the sheet ¼-inch deep. Tilt the tray to distribute the purée evenly.

It takes about two to three days to dry the leather, depending upon the amount of sunlight, humidity, and temperatures. Bring the pan indoors for overnight periods. The leather is done when the layer peels back readily and is sticky to the touch. To remove it from the pan, loosen the edges and peel it back about an inch. Place a sheet of plastic wrap over the leather and begin to roll the layer of fruit in one piece into a loose roll. The fruit roll will keep for months in the refrigerator or up to thirty weeks at room temperature of 70° F or less. For longer storage place it in the freezer.

Banana-Peanut Butter Leather. You will need 3 cups ripe bananas, 1 cup chunky peanut butter, and 1 teaspoon vanilla. Blend the ingredients together, then follow the fruit leather recipe above.

Spiced Apple Leather. You will need 8 medium apples, cored, peeled, and cut in chunks; ½ cup light corn syrup; 2 tablespoons lemon juice, ¼ teaspoon ground cinnamon, and ⅛ teaspoon cloves. Blend together, and follow fruit leather recipe above.

Dried Fruit Cookies

½ cup cooking oil
½ cup honey
2 eggs, slightly beaten
1 teaspoon vanilla
½ cup soy flour
½ cup oat flour
½ cup dry milk powder
½ cup dry wheat germ
½ teaspoon salt
½ cup dried fruit (one or a combination), chopped

With a whisk mix oil and honey until it is well blended. Add eggs and vanilla. Combine and sift together the dry ingredients. Add the fruit to the flour mixture, separating the fruit; then add the wheat germ. Mix together the wet and dry ingredients and drop by spoonfuls onto oiled cookie sheets. Bake at 350° F for 15 minutes. Makes 3 dozen.

2

Solar Greenhouses

One of the most popular applications of home solar energy use is the greenhouse. It is also one of the simplest to understand. For many years greenhouses have been used to collect the sun's heat and energy in order to nurture plant growth. Gradually, an expansion of this basic idea is taking place in the United States. We are seeing the emergence of a wide variety of structures whose purpose is *sunspace*, or space heating. In most climates, heat can be collected and used directly during the day. The excess can be stored for use at night and during cloudy days.

The area of residential-use greenhouses includes everything from passive-solar window enclosures suitable for plant growing in highrise apartments, to total living space such as gardenlike family rooms called *solariums*. The current trend in interior design reflects the sense of well being one derives from living with plants. Since we have taken to surrounding ourselves with an abundance of vegetation, greenhouses are becoming one of the most popular features in new home construction.

You should first decide what the major use of your greenhouse will be—plant production or residential space heating. If the main purpose is to grow plants, the temperature and humidity ranges must be more closely controlled than for the production of heat. For example, direct summer sun in a greenhouse with a southern exposure can easily produce temperatures above 100° F that can kill many plants overnight. When leaf temperatures on plants reach 120° F, cells are killed and the leaves scorch immediately.

In this chapter, I will concentrate on greenhouse or *growing* techniques and *combination* (growing and heating) approaches made possible by the intelligent use of glazing, shading, movable insulation, and *thermal mass* for heat storage. Thermal mass refers to any part of the structure, such as stone or masonry walls or flooring, that can absorb solar heat during the day and release it slowly during the night or on cloudy days. Design considerations and options for construction materials will be presented along with suggestions for growing plants and vegetables in a greenhouse environment.

TYPES OF GREENHOUSES

For this discussion, greenhouses are divided into two structural classifications—freestanding and attached—and three functional categories—sunspace (space heating), commercial/hobby (growing), and combination (growing and heating).

In solar greenhouses, the emphasis is placed on providing an environment for plants using solar energy as a primary heat source. In combination designs, excess heat is used directly by the home or is stored for future use. This is certainly beneficial but is considered a by-product of the greenhouse. Sunspace, or *solariums*, or passive solar design is not covered in great detail here, but several good references are given in the Sources of this book.

Freestanding Designs

Freestanding is just what the name implies. The greenhouse structure stands alone on its own foundation and is not supported by, nor attached to, another structure (such as the house). Because there is no connection to living space, the only function a freestanding greenhouse performs is to grow plants. Virtually all commercial greenhouses are of the freestanding type. Some typical designs are shown in Fig. 2-1. These structures are also referred to in the trade as *glazed* buildings because they consist of a metal (or sometimes wood) framework supporting glass or plastic enclosure panels (glazing). Many of the variations shown are available in hobby sizes that will be of most interest to the average homeowner. A good source for information is the Hobby Greenhouse Association, 45 Shady Drive, Wallingford, CN 06492. They publish a bi-monthly magazine called *The Planter* that includes a swap shop section.

Attached Structures

Space limitations both inside and outside the home, coupled with rising food and fuel bills, have created a strong demand for

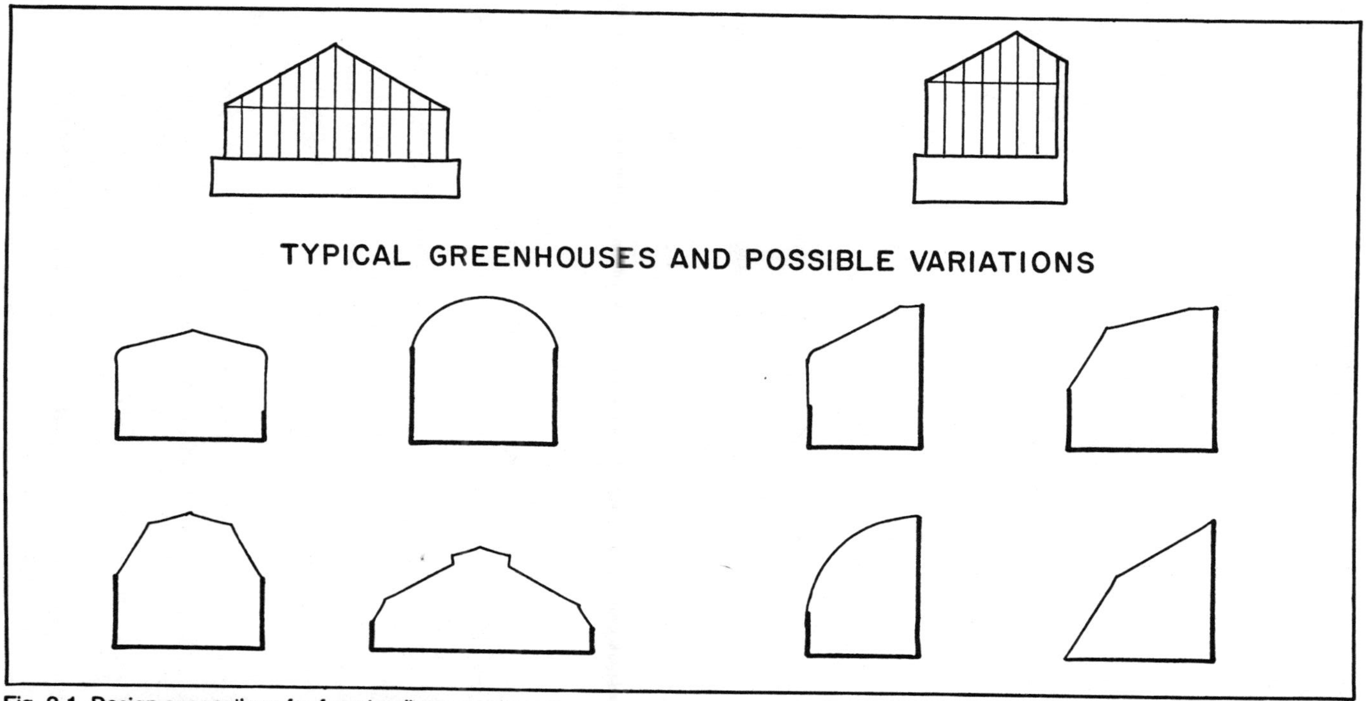

Fig. 2-1. Design suggestions for freestanding greenhouses.

attached, or combination, greenhouses that perform both growing and heating functions. You can extend your living space and provide a garden without supplementary heat. With proper design, this type of structure can also reduce home heating bills by 10 to 30 percent.

Designs vary with the architecture and sitting (orientation) of each home. Figure 2-2 shows one popular approach. Preferably, the greenhouse is added to the south-facing wall of the house. Some latitude is permitted, perhaps ± 20° east or west of south, with proper design adjustments. Primary solar collection is through the large glass window sloped at about 60° from the horizontal. A skylight also provides solar gain and allows more light into the room. The skylight is optional and, if used, a roof vent is strongly suggested to provide summer cooling. All glass surfaces should be double-glazed (two panes of glass) to reduce heat loss. The end walls of the room are generally of solid construction, with no openings, to reduce heat loss in winter.

The masonry floor and thermal mass heat slowly during the day and radiate heat back to the room at night when it is needed. The glass areas can be covered with insulation batts at night to reduce heat losses. As air circulation is required for healthy people and plants, it is necessary to provide a circulation path for the heated air. The heated air can be circulated either through the house when heat is required or through the thermal mass to be stored for later use, or it can be vented outside.

For sunspace, or applications involving space heating only, a design similar to Fig. 2-2 may be used, but more thermal mass should be considered. Basically, you must rely on materials that have a high *thermal mass capacity* (ability to hold heat over a long period) to store excess heat during the day and release it at night.

Those systems where fluid or air is forced to circulate by pumps or blowers are called *active* solar systems. Those providing natural circulation (Fig. 2-3) are called *passive* solar systems. Systems using both natural and forced circulation are called *hybrids*.

The Pueblo Indians have used adobe block to build their homes in the Southwest for years. For their primitive architecture, these homes are quite comfortable. Other solar energy innovators have used 55-gallon steel drums filled with water to construct the south wall of the home. The water heats during a sunny day. In the evening or on cloudy days, an insulated, hinged cover closes over the outside surface to prevent heat loss. At night, the hot drums radiate heat inside the dwelling. In the morning, the outside cover is lowered so that the drums may once again store heat.

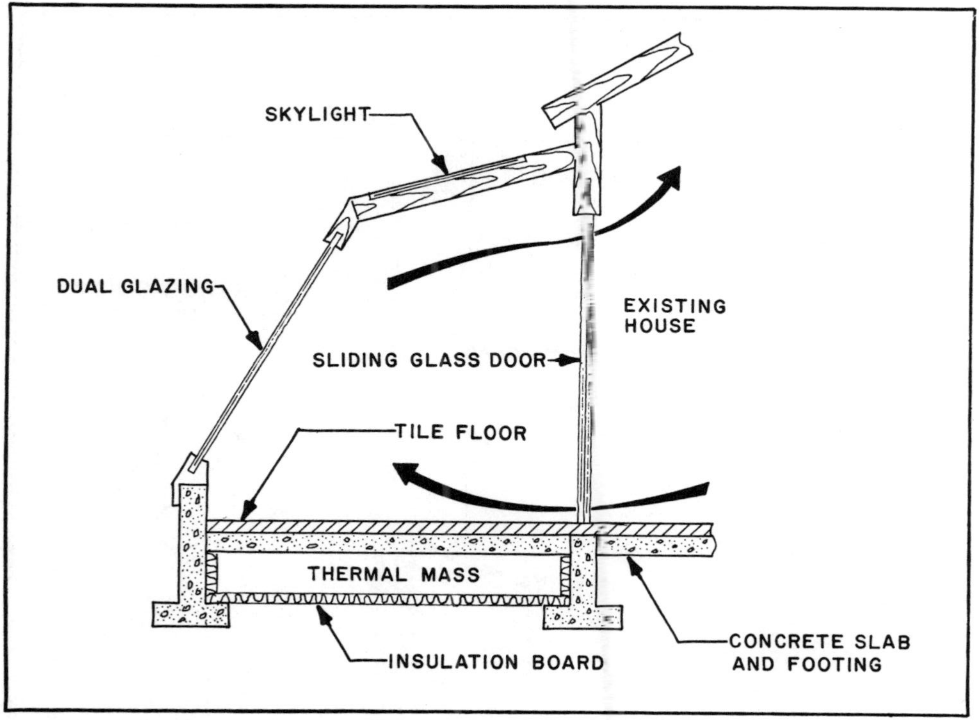

Fig. 2-2. Typical construction and air flow for an attached greenhouse structure.

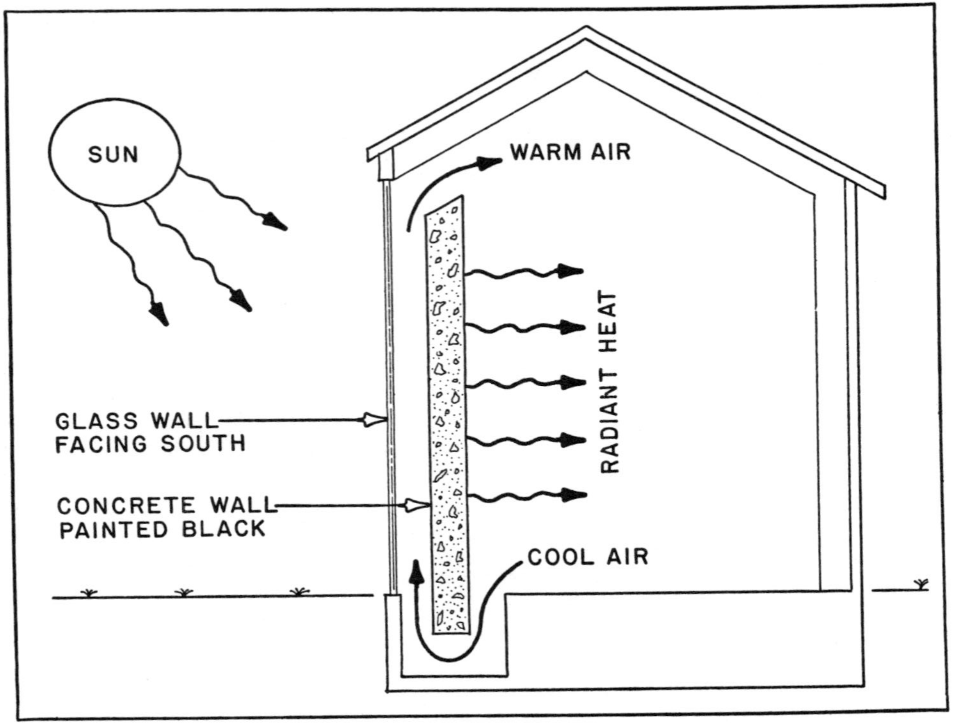

Fig. 2-3. In passive solar heating air circulates by natural convection.

GREENHOUSE THEORY AND ENVIRONMENT

As energy produced by the sun passes through a transparent material such as glass and strikes an object behind it, the light's *wavelength* is changed, trapping this energy in the form of heat. This is called the *greenhouse effect* and is the basis of most solar energy devices (Fig. 2-4). Plants will quickly die in this environment unless humidity, temperature, and ventilation are carefully controlled. On the other hand, if there were drums painted black and filled with water behind the glass, heat could be stored during the day and used to warm surrounding space at night. Each application requires a different environment. Satisfying both in an attached greenhouse, which is a part of the living space, provides the greatest challenge.

First, let's talk about what plants need to grow and why traditional outdoor gardening methods are not compatible with greenhouse growing. The greatest difference is the potential for overheating. Light is generally not a problem out of doors but another factor that must be considered in an enclosed space. If you are considering an attached greenhouse, plants and vegetables should be selected that have growing conditions compatible with the home environment.

Temperature and Ventilation

Most plants will grow well in a greenhouse environment of 68 to 85° F and even tolerate lows of 55 to 60° F at night. Sudden changes and temperature extremes are very harmful. These must be controlled by careful design of the greenhouse structure. If there is a choice, design to the cool side. Plants can recover from a chill, but once they dehydrate from overheating, they will die quickly. Some plants like cooler temperatures; in fact, fruiting plants develop optimum growth where nighttime lows reach 55° F. Carrots and beets like a 45 to 55° F range.

Ventilation (air circulation) must be maintained on a year-round basis for *all* plants. This prevents overheating as well as diseases such as mildew. Most greenhouse designs provide for vents to exhaust hot air outside the structure in the summertime. Once hot air is exhausted, cooler air will flow into the structure by natural convection if intake vents are provided. For attached greenhouses, which are an extension of the living space, excess heat during the day from the greenhouse may be stored or circulated into the house directly as heat is required. It is most important to maintain ventilation and never allow the greenhouse air to become stagnant or stale.

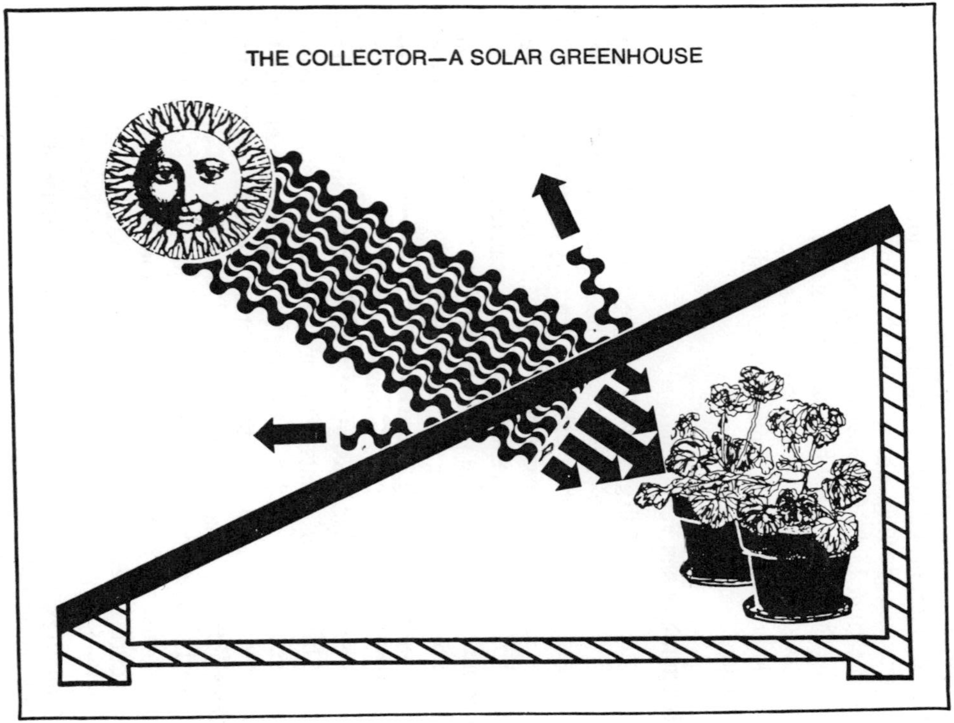

Fig. 2-4. The greenhouse effect. The change in the light's wavelength as it passes through transparent material traps it in the form of heat.

Humidity

The problems caused by high humidity in greenhouse structures outweigh the benefits. A humid environment encourages plant growth but also attracts more insects and creates a breeding ground for plant fungus and bacteria. Generally, a humidity level of 30 to 50 percent is satisfactory for most plants and is also comfortable for people. The best way to measure humidity is with a small ***hygrometer*** in the greenhouse.

For attached greenhouses, it is important to remember that the furnaces used to heat homes in winter dry the air. Thus, you will need additional moisture. If it is very hot, you may need to mist spray plants during the day. Do this before noon, since humidity should be lower at night. Water plants with a medium spray rather than by pouring large amounts in the containers. Overwatering causes root rot and other problems, so keep the greenhouse on the dry side. You should water routinely—varying the quantity to suit your climate and the type of plants.

Condensation commonly found in freestanding greenhouses is generally not a problem with attached units, since air from the structure circulates into the house and lowers humidity in the greenhouse. Moist air will condense out as water at the ***dew point***, the point where the temperature is cool enough to change water vapor to liquid water. This frequently occurs on window surfaces but may be reduced by dual glazing (two panes), recommended for attached greenhouses.

Light and Shade

If excess heat generated during the summer cannot be stored or vented, some shading will be necessary. This may be accomplished by the use of movable panels that also insulate glazed areas at night or by using attractive, selected trellised plantings outside that provide a filtering effect to the incoming sunlight.

Each greenhouse has its individual pattern of light and shade depending upon its orientation and construction. This will vary throughout the year. If sloping glass area faces south (see Fig. 2-2) the top of the north wall next to the sliding glass doors will be shadier in summer when the sun is higher in the sky. This would be a good location for shade loving plants such as berries and sprouts. If the roof skylight were eliminated, shade would, of course, be increased. Once design of the structure is complete, you can decide the best location for plants selected. Winter vegetables need medium light. Summer vegetables, especially peppers and melons, require a lot of light. Some plants, like citrus, need a dormant period. Remember that plants grow *toward* the light. This is called the *phototropic effect*. Plants can be fooled if you place a reflector (such as aluminum foil) behind them to keep them growing straight.

DESIGN CONSIDERATIONS

Although most of the specifics in this section pertain to attached combination greenhouses, the glazing details illustrated may also be considered for freestanding structures.

Location

Attached greenhouses can be a valuable asset in most parts of the United States. The heat-gain, heat-loss, and heat-storage requirements of the structure will vary considerably as a function of local weather conditions, quality of construction, and design. If you are contracting or self-building, the first step is to survey your site for suitability. Look for a location on the south side of the house that is relatively shade-free during winter months.

Most homes will not have an ideal location, so you may consider an orientation that is within 20° east or west of true south without an excessive penalty. Notice I said *true* south. The reading on a magnetic compass should be corrected for your location by referring to an *isogonic chart* or by calling the city engineer's office and asking for the magnetic declination for your area. For example, in Boston, Massachusetts, magnetic north lies 15° west of true north. True south, therefore, is at 180 + 15° or 195° on the compass. For a rough approximation: consider New York City and points east to have true south at 195° compass reading, cities between New York City and Denver, at 180° and cities west of Denver, at 165°.

A quick and easy way to estimate shading effects of nearby trees or buildings is to use a magnetic compass and the Sun Locator, an inexpensive cardboard device available through *Popular Science Magazine* (Fig. 2-5). Take a sighting from the lowest point where greenhouse glazing will be installed and sight along the dotted line while holding the locator in a level position. You can then estimate the lowest position of the sun at various times of the day and determine if any objects will be in the path. Deciduous trees may be acceptable depending upon the amount of summer sun passing through the foliage. If these trees are in the sun's path, be sure to plan for growth, topping, or pruning.

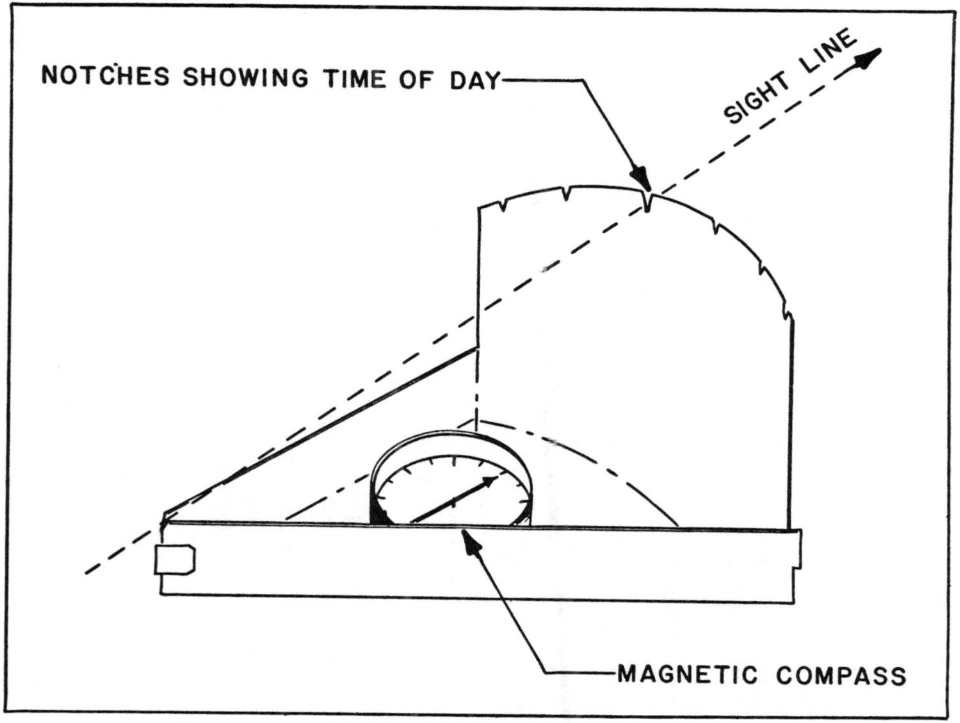

Fig. 2-5. The *Popular Science* Sun Locator is one method of estimating shading effects.

If you have a choice once orientation and shading factors have been satisfied, plan a location along the south wall where the greenhouse will cover an existing door or window for visibility and increased heat gain to the house. A sliding glass door opening onto a patio area is a popular location.

Size

The physical size of an attached greenhouse is generally determined by the space available, architectural considerations, and the budget. The only basic rule is to maximize the length over width. Typical smaller sizes are 8 by 10 feet; larger ones are 8 to 10 feet deep by 20 feet long.

The slope of the terrain next to the house is a factor in deciding if the greenhouse floor will be at the same level as the floor level of the house or if a "step-down" variety is the best choice. Other considerations involve the existing roof overhang for single story homes and the glazing angle selected for the best heat balance. It is important to remember that in a step-down design, a cold air trap is naturally formed at the lowest point. A circulating fan can be used, however, to minimize the cold air pocket. Most existing patio slabs are below floor level. You might consider adding a thermal mass on top of the slabs and pouring a new slab, or using loose brick at floor level (see Fig. 2-2). A foundation around the outside perimeter would, of course, be required.

Heat Balance and Storage

Once the location and physical size of the greenhouse has been selected, the next step is to determine the area and orientation of south-facing glass that will be used. I recommend that you consider only dual-pane insulating glass. First, determine the heat *gain* through the glass; then subtract the heat *losses* through the structure. This equals the ***net*** heat gain available to heat plants and people. There is no heat gain through the glass at night, of course, only loss. An insulating curtain or movable batts can be placed over the windows to retain the heat, which is supplied from storage.

As mentioned earlier, overheating during the day must be prevented, because this will kill plants quickly. This is where storage in the thermal mass of the structure, ventilation, and air circulation to the house must be considered.

Table 2-1 will assist you in selecting various heat storage options. Two common latitudes were chosen, 32° and 40°, and three glazing angles (angle between the glass and horizontal). Surfaces that are perpendicular to the angle of the sun provide the greatest gain. This is apparent in the column for 90° (or vertical) glazing. As the sun is lower in the sky during winter, more energy passes through at this angle in the winter than during summer. You will also see an increase at the more northern latitude (40° versus 32°).

Here is an example of how the table may be used: the location is New York City, 40° north latitude, and a glazing angle of 50° has been selected. The gross solar gain in March is 2010 Btu per square foot per day. The heat loss through double-pane insulating glass is approximately 0.65 Btu per hour for each degree Fahrenheit temperature difference between the greenhouse and the outside air. Assuming this difference is 40 degrees:

heat loss during day = 0.65 × 8 (hours) × 40 (degrees)
= 208 Btu/sq ft

net heat gain = 2010 – 208
through glass = 1802 Btu/sq ft-day

This is the amount of heat that must be stored or used inside the house. Storage is the first consideration.

The most common approaches to providing thermal mass storage are solid masonry (brick, cement, etc.) or metal containers filled with water. Masonry up to 8 inches thick may be expected to store 15 Btu, per degree F, per square foot of surface exposed to the sun. Water containers with a minimum surface area of 1 square foot per gallon stored, can hold 8 Btu per degree F, per gallon. With the

Table 2-1. Solar Gain through South-Facing Dual-Glazed Insulating Glass (Btu/Sq. Ft. per Day).

Latitude	32°N			40°N		
Glazing Angle	40°	50°	90°	40°	50°	90°
Month						
J	1864	1906	1566	1654	1677	1519
F	2064	2043	1447	1903	1938	1522
M	2075	1976	1123	2050	2010	1306
A	1941	1755	672	2042	1908	899
M	1816	1573	413	1992	1795	637
J	1751	1487	326	1957	1737	537
J	1786	1543	403	1962	1765	618
A	1887	1702	648	1987	1852	861
S	1992	1896	1079	1961	1920	1246
O	1982	1964	1397	1813	1846	1456
N	1834	1874	1533	1565	1646	1484
D	1774	1836	1579	1438	1531	1448
Annual	22,766	21,555	12,186	22,324	21,625	13,533

40° differential in the previous example, you could absorb 600 Btu/sq ft (15 × 40) by using a masonry floor and back wall in the greenhouse. The remaining 1202 Btu could be stored in water containers or vented into the house during the winter and into an attic or out of doors during the summer. A small window-type exhaust fan with a capacity of about 5 cubic feet per minute for each square foot of glazing used is a popular choice. Other design options are possible (Fig. 2-6). There are no hard and fast rules on how storage and ventilation must be used, and there is flexibility depending upon space requirements, home heating needs, and nighttime low temperatures expected.

End Walls and Skylights

For south-facing greenhouses in cold climates, I prefer to eliminate all glazing on the end walls (east and west) and the use of skylight. The east and west walls contribute a minimum solar gain and are areas where a significant amount of heat can be lost in winter if they are glazed. Additionally, glazing would add to construction cost and require the installation of insulation batts at night to reduce heat loss. Roof glazing or skylighting is not very efficient from a heat collection standpoint but may be desirable to improve the appearance of the room or add more light. If you decide to put glazing on the roof, you should provide shading control in the summer to prevent overheating and insulation during winter to prevent heat loss. Greenhouses that do not face true south but perhaps 20° east or west of south, provide a special case where end wall glazing may be considered.

CONSTRUCTION DETAILS

Construction requirements will vary due to architecture, climate, soil conditions, etc. In addition, each community has building codes to satisfy before a building permit may be obtained. For these reasons, a set of plans cannot be provided. Instead, I have included notes on some common construction techniques that you may find helpful in designing your own structure or discussing it with your contractor.

Perimeter Wall Foundation

The foundation of the greenhouse structure must support the "dead" load or weight of the building by distributing the weight over the entire perimeter in such a manner to prevent non-uniform settlement or "sinking." Additionally, it must support "live" loads

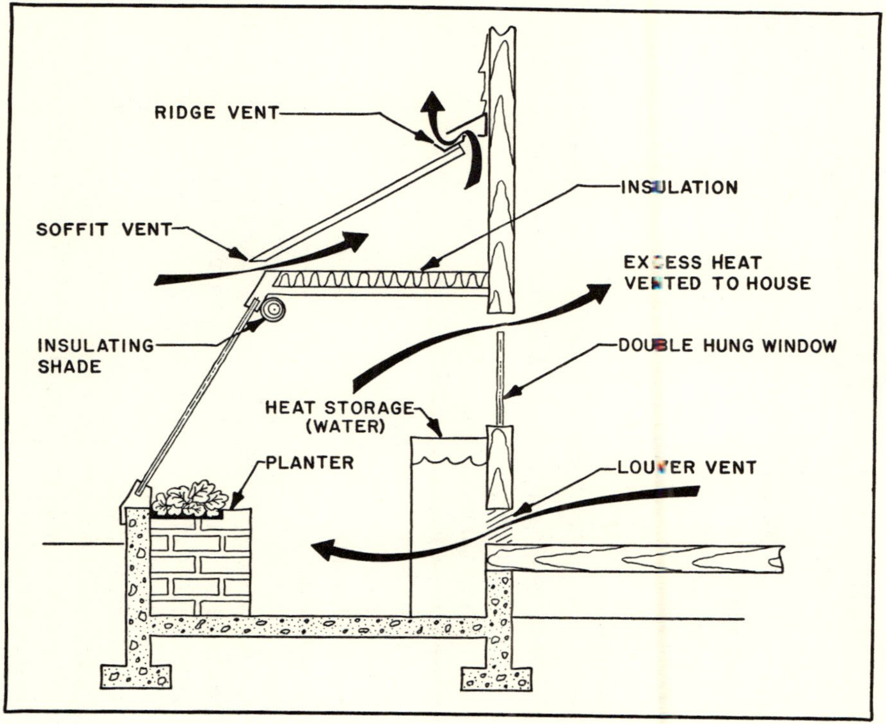

Fig. 2-6. Design options for attached greenhouses. The arrows show air circulation inside the structure and the attic.

such as rain, snow, and wind forces that may tend to lift the structure out of the ground. One critical factor in selecting a foundation design is the type and bearing capacity of the soil. Soil retains moisture. During the winter/spring season in cold climates, the freeze/thaw cycle causes the earth to heave. This could cause severe damage if the footing (the part resting directly on the soil) of the foundation is not placed below the frost line (Fig. 2-7).

The foundation is generally constructed of poured concrete or concrete block, depending upon the availability of either. The cross section of the foundation wall resembles an inverted "T." The footing or bearing surface is about twice the width of the foundation wall with the same thickness. In addition, in order to provide a greater area over which the load may be distributed, this design provides an anchor against uplift and racking forces if the structure is properly attached to the foundation wall. To make this connection, a wooden sill plate is set on a mortar bed or fiberboard base, sealed, and attached to the foundation wall with anchor bolts.

Foundations should be protected against water and moisture. To do this slope or grade the surface *away* from the wall to drain the surface water. Vapor barriers both above and below grade are suggested. Two coats of an asphalt base coating, silicone, or plastic may be used.

Insulation

The perimeter wall foundation just discussed must be insulated from the sill plate to the top of the footing. I recommend using Styrofoam SM-type insulation boards manufactured by Dow Chemical Co. (or equal) in a 1-inch thickness. This material is a closed-cell polystyrene foam supplied in rigid board form and used extensively in walls and foundations. You can place it either inside or outside the wall depending on your specific design. The closed-cell structure has a good insulation value and will not absorb water.

If you use an attic structure (see Fig. 2-6), the ceiling should have a minimum R-factor of 30, or about 8 inches of fiberglass (1 inch thickness equals about R-4). For end walls, I suggest an R-value of 16. These values are not absolute and may be reduced depending upon where you live. Maps showing recommended insulation levels for walls and ceilings in six U.S. climatic zones are available from fiberglass manufacturers. If you use blanket- or batt-type insulation with a moisture barrier (foil faced, for example), be sure to install it with the barrier facing toward the inside of the greenhouse.

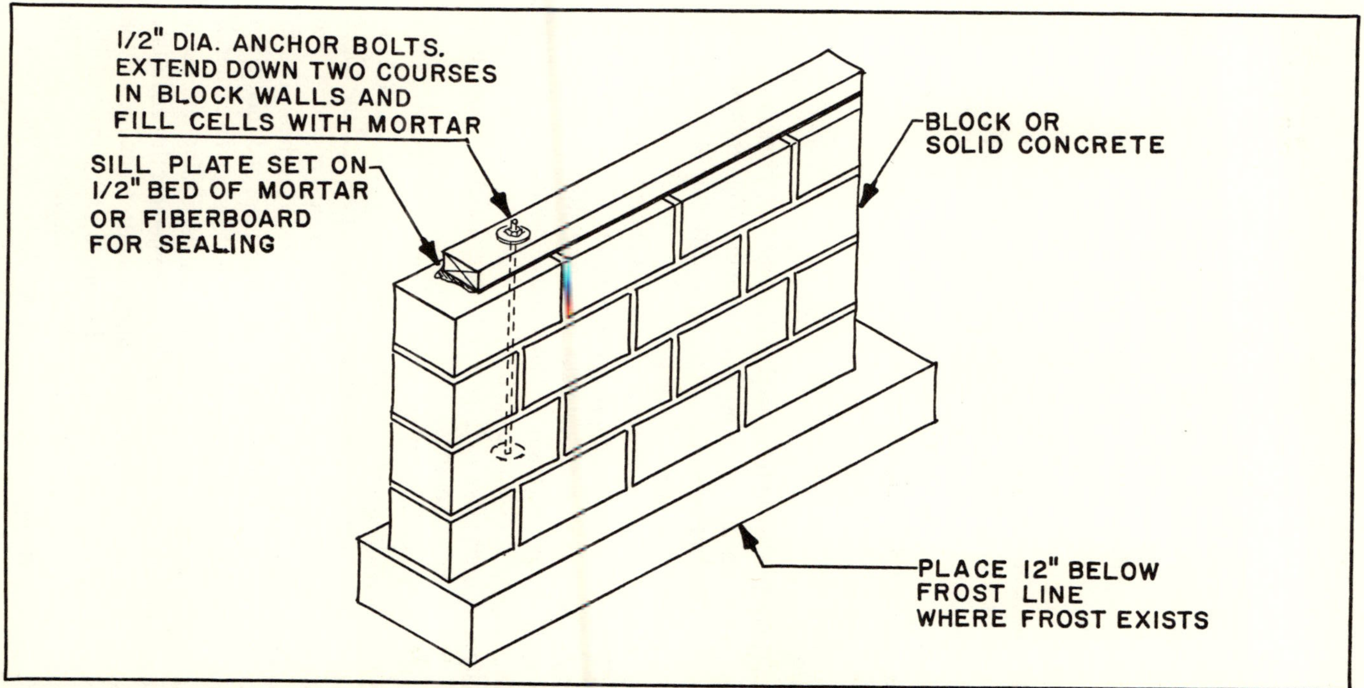

Fig. 2-7. Typical construction of perimeter wall foundation. Use reinforcing bars as necessary to meet local conditions.

The 1-inch-thick foam board material used on the foundation will also provide excellent insulation for the greenhouse windows at night if you care to build removable batts to cover these areas. An insulating curtain or window shade device is easier to manipulate but will not provide the insulating value of the Styrofoam board.

Caulking and sealing is an important part of the insulating process, for air is a major cause of heat loss. Sealing between the sill plate and foundation is most important. Other common areas where infiltration occurs are corners, the intersections of window and door trim, and around pipes, vents, and similar places that penetrate the outside walls of the room. Weather stripping outside windows or doors is a *must*. Silicone caulking may be more expensive than butyl, but it won't crack in cold weather. It is the best thing to use in places where there is expansion and contraction. You may use the more economical butyl in areas where cracking is less likely to occur.

Glazing

Glazing is a broad term used to describe how glass or plastic materials are secured in a prepared opening. This includes the selection of sealants, fasteners, flashing, and framing that provides a weather-resistant seal and compensates for expansion and contraction of the different materials being assembled (Fig. 2-8).

There are two design recommendations that I feel quite strongly about: first, use dual-pane insulating glass in colder climates and second, carefully consider the safety hazards of using glass in sloping or overhead applications. Most building codes now specify tempered glass in large openings and plastic material in skylights for safety reasons.

Insulating glass, by my description, is a sandwich of two glass panes separated by a metal spacer. A desiccant, it placed inside the spacer to *absorb* (remove) water and solvent vapors, and the edges of the glass are then sealed. The most economical approach is to design around the standard patio door size (36 by 76 inches). The air space between the panes is generally 3/16 inch; and this is the figure I used in developing the example of heat loss given earlier. If your climate is mild enough, consider single glazing and use a heat-loss factor of 1.1. In your calculations remember that nighttime losses may be considerable and you might have to depend heavily on the use of insulation batts in winter. Tempered glass is produced by reheating annealed glass, then quenching (or chilling) it with an air blast. This technique results in an extremely strong, impact-

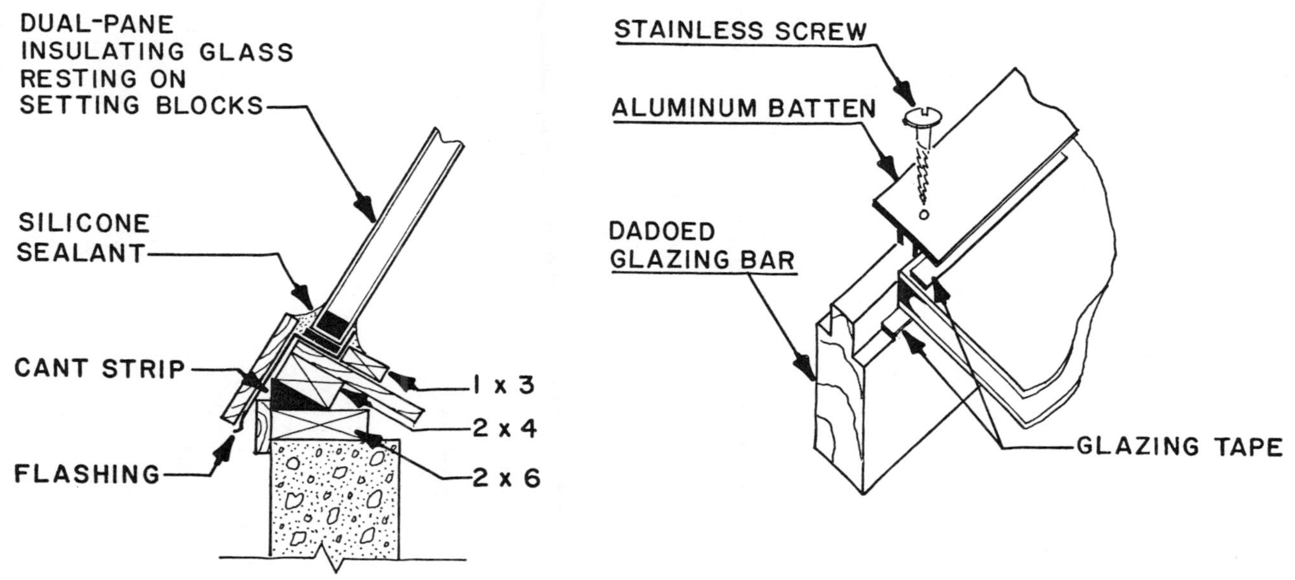

Fig. 2-8. Glazing detail. This typical design for sloping surfaces and sill plate attachment uses insulating glass.

resistant product. When it does break, it does not crack like annealed glass. It literally explodes into hundreds of small, marblelike particles with relatively smooth edges. You can rub a handful of these particles between your hands without getting cut. For these strength and safety reasons tempered glass is generally used for automobile windshields and is specified for sliding patio doors. Because tempered glass cannot be cut after processing, you must buy it to size. For roof glazing, a double-skinned SDP acrylic plastic sheet material, manufactured by CY/RO Industries, is my choice. The material has internal ribs separating two 0.067-inch skins and is about 5/8-inch thick overall. In clear form, the light transmittance is 83 percent, and the heat loss factor is 0.58. This compares favorably to insulating glass. For technical details concerning such factors as joining, loading, and durability, contact the Technical Service Department of CY/RO Industries (697 Route 46, Clifton, NJ 07015).

Figure 2-8 provides typical glazing details for a vertical or sloping greenhouse wall using tempered insulating glass. Glazing tape is a nonabsorbing, close-cell material with an adhesive backing that is available in rolls. Some trade names are Silglaze, Polyskin, and Strip-N-Stick. It is used for the compression type glazing shown. Setting blocks are hard neoprene spacers used to position the glass at the base of the opening. Aluminum battens with stiffening ribs held in place by stainless steel screws provide a good combination.

In addition to being weather tight, the design must satisfy building code requirements for glass under wind-load pressure. This, of course, varies by geographic region. Proper edge support and clearance for the glass in the wooden opening must also be maintained. An excellent source for more detail is the "Flat Glass Marketing Association Glazing Manual." Copies may be obtained at a nominal charge by contacting FGMA, White Lakes Professional Building, 3310 Harrison, Topeka, KA 66611. Here are a few general recommendations to keep in mind:

- The glass should, in effect, "float in the opening."
- Avoid contact between glass and metal, or any hard object.
- Use setting blocks on all glass over 6 square feet in area. Use two setting blocks of neoprene 70 to 90 durometer, installed at quarter points. The blocks should be wide enough so that the glass will not slip off under heavy vibration, wind load, etc. The length of the blocks should be a minimum of 3 inches.
- Design the glazing system so that building movements, inherent with wind loads or thermal effects, are not transferred to the glass.

- Attach the frame to the structure so that it will withstand design wind load and comply with the building code. Most codes state that the deflection of the frame should not exceed 1/175 of its span or 3/4 inch, whichever is least.
- Design the glazing system so that moisture does not accumulate in the glazing channel at the sill for extended periods.
- Coat the glazing channel and all sealed surfaces of wood or carbon steel with a paint primer.

Ventilation and Air Circulation

The importance of good ventilation and air movement has been discussed. You can rely on a small fan to move the air, natural convection, or both. The basic rule is to allow an entry point for cool air in the lower portion of the greenhouse, and a vent for excess heat

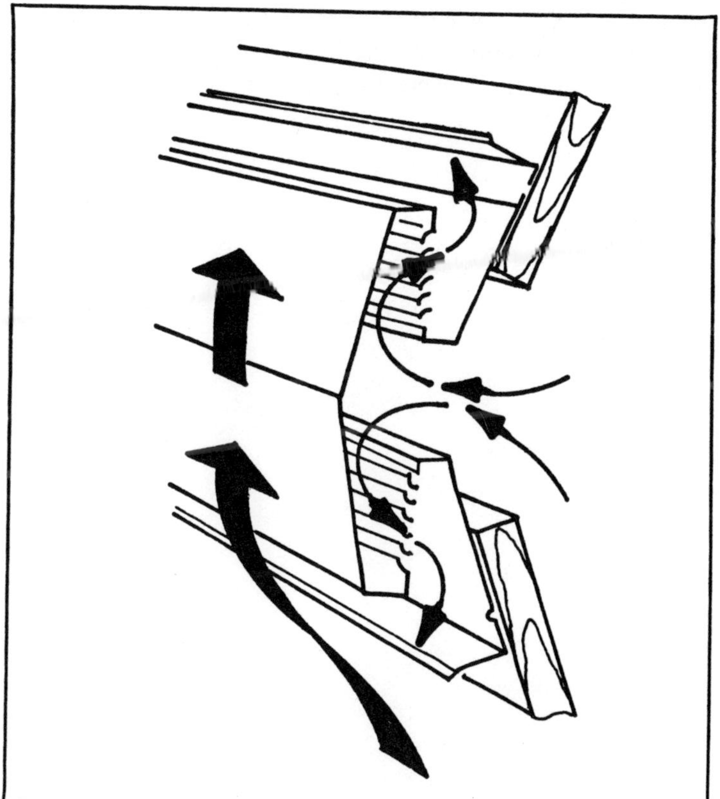

Fig. 2-9. Full ridge vent detail showing air movement. A half section is used to intersect a vertical wall.

to the outside in summer and into the home in winter. In most cases you will have an access door (see Fig. 2-2) or window (see Fig. 2-6) that can be adjusted to provide circulation.

If you extend the existing roofline or build other designs where an attic is used, consideration must be given to proper attic ventilation. Fiberglass ceiling insulation (see Fig. 2-6) keeps the room warmer in winter and cooler in summer, but heat buildup in the attic in summer and moisture condensation in winter can cause problems that proper attic ventilation can eliminate. Control is accomplished using a soffit and ridge vent (Fig. 2-9).

Warm air rises naturally and will escape through the perforated ridge vent. As it escapes it will be replaced by cooler air, that enters through the soffit. This natural effect is enhanced when wind blows over the roof. This creates a negative pressure at the ridge which tends to pull the air from the attic—similar to the lift forces on an airplane wing. Sheet metal vents are readily available, designed to prevent rain and snow from entering the attic when properly flashed and installed with the louvers (or perforations) in a horizontal position.

3

Electricity from the Sun

Many people continue to confuse the way solar energy is used and the terms associated with its different applications. This is especially true when distinguishing between *solar-thermal*, or using sunlight to generate *heat*, and *solar-electric*, which relies on light for the direct generation of *electricity*. The field of solar-electrics is commonly referred to as *photovoltaics*. Units of light energy are called *photons*. The *volt* is a force which causes current to flow. Thus, when a device called a *solar cell* converts light energy into electrical energy, we call it the *photovoltaic effect*.

Perhaps the most well-known application of the photovoltaic effect (electricity generated directly from sunlight) is the exposure meter used on cameras. The amount of electricity produced is too small to power even a light bulb. The same principle of changing light energy directly into electricity is being used in studies to produce more efficient devices for larger loads.

The most commonly used device for converting sunlight directly to electricity is the photovoltaic or solar cell that was developed by the Bell Telephone Laboratories in the early 1940s to improve telephone communications. The United States space program in the 1960s, however, provided the money to further develop these cells. They were used to provide electric power to space vehicles and satellites. Generating electricity in this manner was mandatory in the spacecraft. On earth, where there are other energy sources available, solar cells must offer advantages commensurate with their cost if they are to be sold in large volume to the public. Solar cells do offer a number of positive features: they have no moving parts to wear out; are quiet, safe, and nonpolluting, and if protected from moisture and excessive heat, can outlast the people who use them. The following section describes the solar cell in more detail and explains how they work, how they are manufactured, and what you can expect from them.

SOLAR CELLS

Solar cells are constructed of various materials called *semiconductors*. A semiconductor requires heat or light to conduct electricity. The vast majority of cells manufactured today are made from silicon, the second most prevalent element in the earth's crust. Silicon is the shiny substance you see in grains of sand and is also the material from which transistors are made.

The cell is assembled from two thin wafers of silicon to which impurities have been added to create negative (N) and positive (P) charged layers. The interface between the wafers is called the P-N junction.

In Fig. 3-1, sunlight strikes the top of the cell and causes *electrons* (a charge of negative electricity) to be dislodged within the crystal. These tiny charged particles migrate across the P-N junction creating an electric field across it. When wires are attached to the top and bottom cell surfaces, a path is formed for current to flow through to the load (motor, light, etc.), completing the electric circuit. Current will continue to flow as long as the light energy striking the cell is strong enough to overcome the resistance of the load. As electrons flow in one direction only, the current produced is called *direct current* or dc.

Most electrical appliances used in the home operate on *alternating* current or ac and require a device called an *inverter* to convert the dc output of a solar cell to ac. Some appliances, such as a transistor radio or portable television set, can operate directly from solar cells. A small TV might require 15 to 20 watts to operate (watts = volts × amps)—or an approximate area of 2 square feet of properly matched and connected solar cells. Figure 3-2 shows such an application using a 20-watt (peak) solar module by Motorola.

How Cells Are Made

Cell fabrication begins with the purification process. Crystals 3 to 4 inches in diameter and several feet long are "grown" by the rotation of a single pure "seed" crystal in a bath of molten silicon.

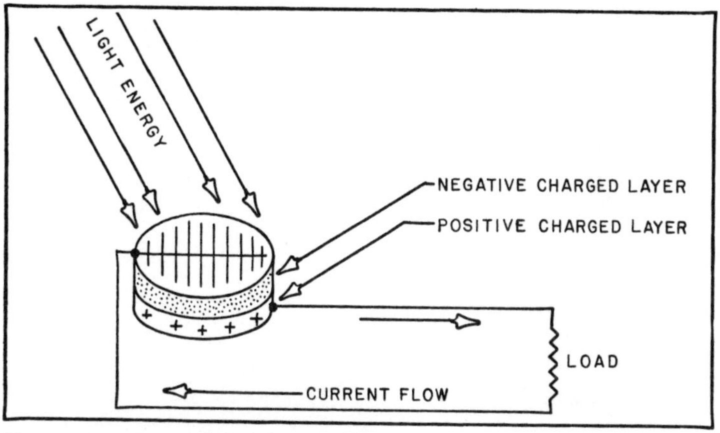

Fig. 3-1. Principle of photovoltaic conversion.

Once the crystal is withdrawn from the bath, it is trimmed and sliced like a salami into discs about .02-inch thick. A small amount of boron is added to the molten bath, giving each of these discs a positive electrical charge. They become the bottom layer of a photovoltaic cell. To form the top or negative layer, the disc is heated in a phosphorus atmosphere where a few atoms of this element diffuse into the surface. This process is called *doping*. Next, the positive and negative layers are joined and the interface is called the P-N junction. Electrical contacts are applied to both the top and bottom of the assembly—a comblike metallic contact grid to the top or negative surface, and a conductive material to the entire bottom, or positive, surface. Figure 3-3 is a schematic representation of the assembly. The last step includes hermetically sealing the unit to keep out rain, hail, sea spray, dirt, and moisture, thus prolonging the service life of the cell.

Although most cells are round, they may be cut into various shapes to increase the density within a solar panel. This provides cost savings and a choice in electrical output. Other methods of manufacturing silicon cells under study promise to reduce cost by the elimination of the sawing, or slicing, step. Growing thin flat crystals in either ribbon or sheet form may produce more useful material, as slicing the wafers from a log generates about 50 percent scrap in the form of "sawdust."

Performance of Cells

The amount of power (watts) produced by an individual cell is determined by several factors. Aside from the quality of materials

Fig. 3-2. Solar module (20-watt) powering TV set (courtesy, Motorola).

used in construction and proprietary assembly features, solar cells can be expected to produce approximately ½-volt regardless of the area exposed. On the other hand, the amount of current increases with both the area and the intensity and quality of light energy striking its surface. Nearly half of this energy is in the wrong form to be useful. For example, sunlight in the infrared region produces heat in the cell that lowers efficiency and electrical output. Considering physical limitations, the maximum theoretical efficiency of a silicon cell is about 21 percent. In practice, these devices operate in the range of 10 to 14 percent, producing about one *peak-watt* of

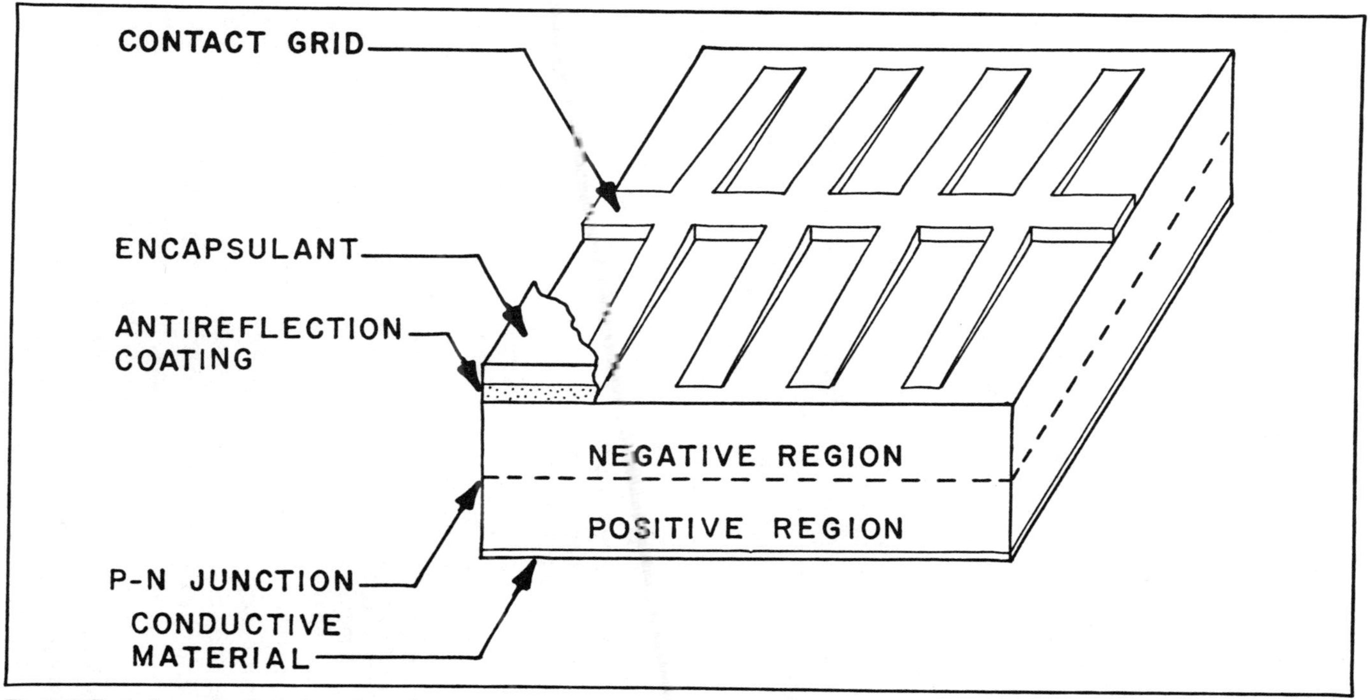

Fig. 3-3. Typical construction of a solar cell.

power from a 4-inch cell. A peak-watt is the amount of power that can be generated at high noon, on a clear day at a temperature of 77° F, with the cell *normal* (or perpendicular) to the sun's rays, that is, the amount of power obtained under ideal conditions. Because the sun's position and intensity changes hourly and seasonally, it is not possible to expect this peak watt output in practice. With the cells in a fixed position, it has been estimated that the average output during daylight hours is about 30 percent less than if allowed to "track" the sun.

There are various manufacturing techniques that can be used to improve cell efficiency. Some producers etch or "texturize" the top silicon wafer to increase the surface area and light absorption capability for a given cell diameter. Also, a proprietary antireflection coating is applied to most cells. Such coatings can improve efficiency 2 to 3 percent by reflecting the unwanted infrared radiation and absorbing more light in desirable wavelengths.

Aside from obvious shading from trees, structures, and the like, the electrical contact grid on the surface of the cell blocks sunlight. Therefore, it is kept as small as possible without excessive resistance to the flow of electrons. Once a grid design has been selected by the cell manufacturer, this pattern is printed on the top surface using a standard photographic resist process. The antireflective coating is etched away where electrical contacts are desired. A highly pure and moisture resistant conductive material is then deposited over the etched areas to produce the contact grid. You will note in Fig. 3-3 that the contact grid is heavier near the center of the cell where the current load is greater.

The voltage and current produced by a cell varies as a function of the load imposed, in addition to the intensity of sunlight. To measure this, current versus voltage curves are developed for cells based upon sunlight of a fixed intensity. The value of 100 mw/cm^2 (one hundred milliwatts per square centimeter), or in English units about 1 kilowatt per square yard, is commonly used. This reference value is also called one ***sun*** and represents about the normal amount of solar energy falling on a 1 square centimeter surface at noon on a clear day. Figure 3-4 shows the current-voltage curve for a typical cell. With no load applied, the circuit is open and no current will flow. This is point A on the curve. If a load were applied that had little or no resistance, such as a single wire, a short circuit is produced and there is no voltage. This is point B on the curve. Since there can be no power output if either current or voltage is zero, the best operating point would be at the "knee" of the curve where the product of current and voltage is maximum. This is shown as point C.

To further define the cell's performance, other light intensities should be shown. Figure 3-5 illustrates the electrical output expected at various light levels. It is important to note that the voltage produced remains about 0.5 at the knee of each curve even though the light intensity increases. The size of the cell, therefore, is important in determining current capabilities, and the number of cells (in series) provide voltage increases.

Another factor affecting performance is the temperature at which the cell operates. The maximum power output increases about 10 percent at 40° F and decreases approximately 20 percent

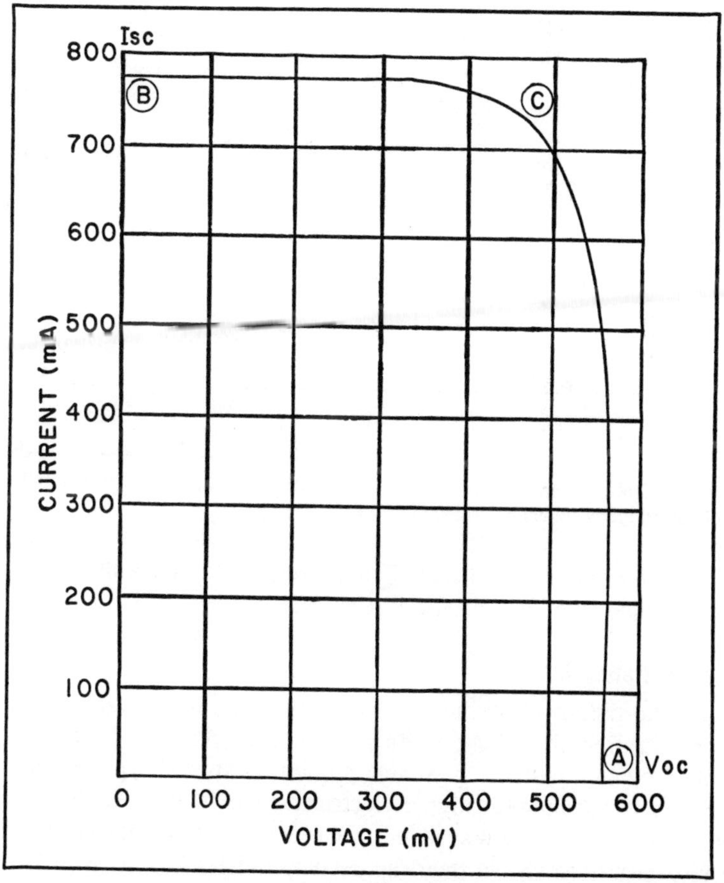

Fig. 3-4. Current-voltage curve for a typical solar cell. Desired operation is at point C.

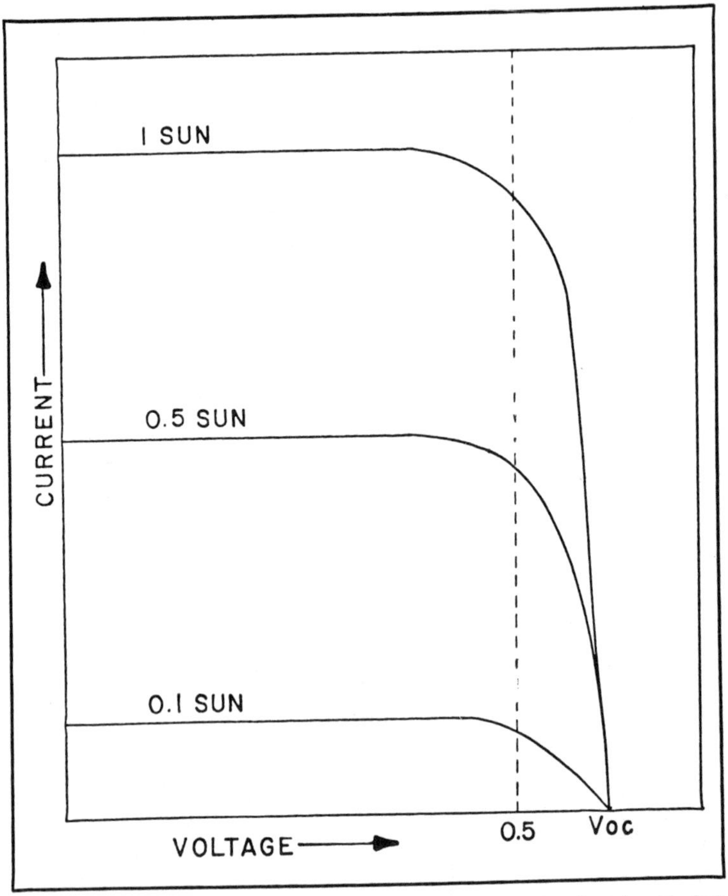

Fig. 3-5. Effect of various light levels. Current increases with solar intensity, but voltage remains relatively constant.

at 160° F from the baseline of 77° F. More precise values should be determined from cell manufacturers' data.

Concentrating the Sun

The output of a silicon solar cell may be increased by the concentration of sunlight on its surface. The additional heat produced by this method, however, requires a special type of cell capable of handling the higher electrical currents and heat generated. Such cells are now available. Tests conducted at Arizona State University have shown they can develop efficiencies of 18 to 19 percent under illumination levels of 30 to 300 suns. This 19 percent efficiency might not sound like much, but it is quite an improvement when compared to the 10 percent normally associated with silicon cells. Concentrating or focusing the sun, therefore, can substantially reduce the number of cells required to provide a given power output. The bottom line, of course, is the cost per watt. This will depend upon the premium for the special cell, type of concentrator selected, tracking mechanism, and the use of the heat generated, along with other factors.

There are two basic methods used to concentrate sunlight: one is to reflect the light from a shiny surface to the cell by using either a line focusing parabolic through reflector or a point focus reflector similar to Fig. 1-9. The other is to *refract* light from a large area to a small one using a lens. Figure 3-6 shows an installation using fresnel-type lenses covering 2¼-inch concentrator cells provided by Applied Solar Energy Corporation. These cells are capable of operating efficiently at concentrations up to 100 suns. This particular solar system using 40x concentration is part of a joint U.S. Government project to power two rural villages in Saudi Arabia.

A number of designs have been evaluated to make use of the unwanted heat produced by concentrating-type photovoltaic systems. A common approach is to use a metal tubeway behind the cell. A liquid flows through the tube to remove heat at a given rate. The heated fluid, which may be water, is then used for normal *solar-thermal* applications such as washing, bathing, and space heating. It is possible that these combined uses will help reduce the cost problem encountered when considering solar cells for residential applications.

PUTTING THE PIECES TOGETHER

A single cell has little practical value due to its limited output and lack of storage capability. The next step is the assembly of cells and other hardware into an effective system. This requires the careful matching of the load and supply with a provision for storage. Other criteria may be established such as reliability, low maintenance, useful life, protection from lightning, and other environmental hazards.

Modules and Arrays

To achieve a more useful wattage, or power output, it is necessary to combine several cells in a series combination, parallel combination or series/parallel combination. Adding cells in series increase the voltage, and the number of parallel cells determines the current capacity.

Fig. 3-6. Concentrator system using fresnel lens (courtesy, Applied Solar Energy Corp.).

These assemblies are commonly referred to as *panels* or *modules* and are available commercially from several companies. The panels shown in Fig. 3-7 are manufactured by Applied Solar Energy Corporation. Each includes an assembly of photovoltaic cells encapsulated between two sheets of polyvinyl butyral (PVB) backed with white Tedlar and laminated to special solar glass. A heavy duty anodized aluminum frame provides structural rigidity.

Although it is possible for the home experimenter to buy individual cells and connect them into a module, the commercially

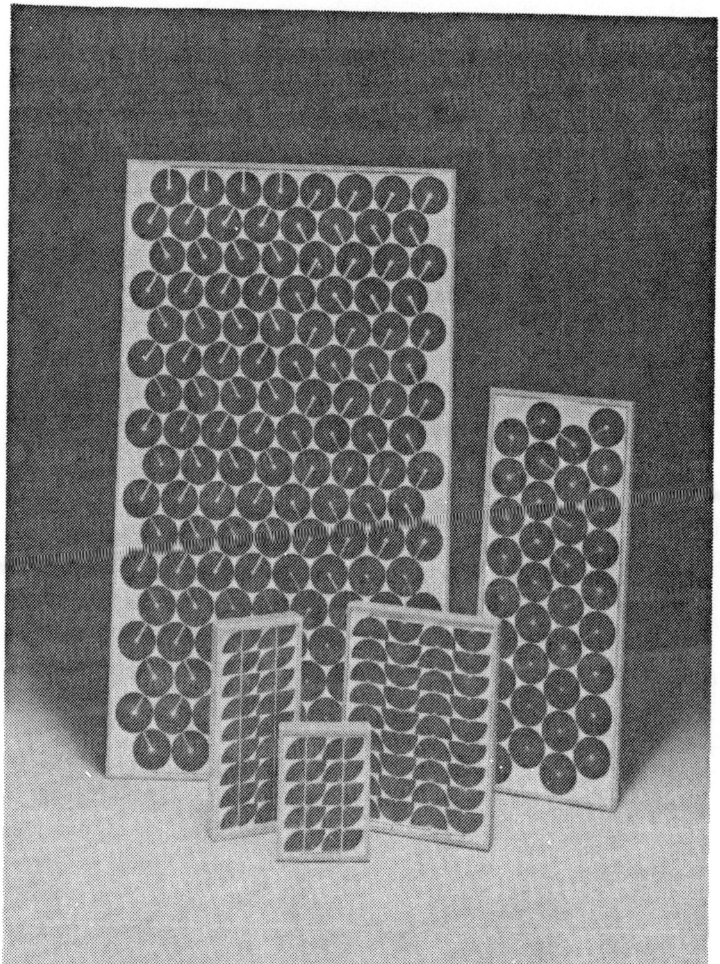

Fig. 3-7. Photovoltaic panels come in different sizes and output. Panels in foreground show ½- and ¼-cell configurations to save space.

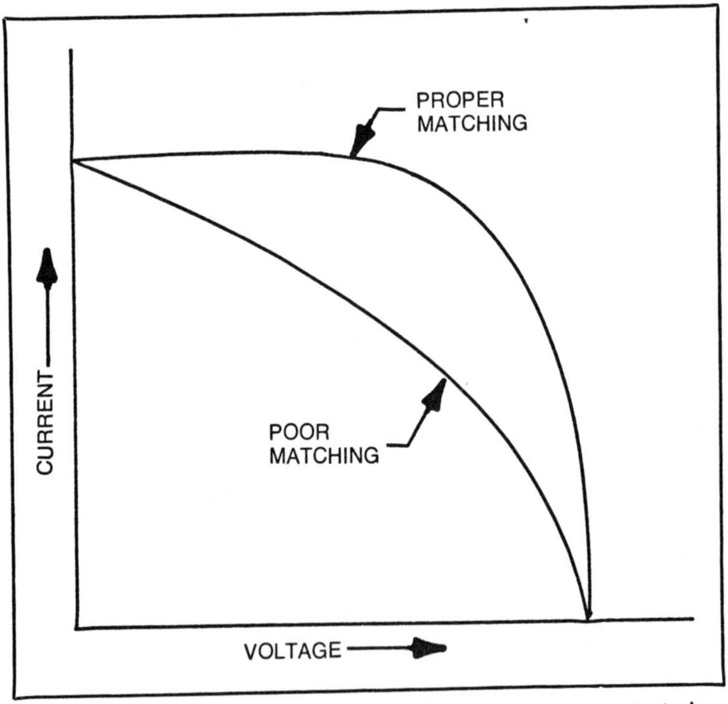

Fig. 3-8. Cells must be matched electrically to insure maximum output when constructing a module.

available units will generally have higher output, as individual cell testing and matching is performed. The output (product of current and voltage) can fall off substantially without proper matching (Fig. 3-8). In addition to electrically matching cells, it is most important to ensure that cells connected in series in a module receive roughly equal amounts of sunlight, because the cell with the lowest illumination may, in series configuration, control the current for the entire string of cells. A shadow cast on one cell, for example, could "open circuit" the module just as one faulty bulb in a series string of Christmas tree lights causes all the bulbs to go out. A summary of electrical characteristics for several commercial modules manufactured by Applied Solar Energy Corporation are shown in Table 3-1. These models use 3-inch-diameter cells with an output ranging from 12 to 88 watts.

For applications requiring greater power output, two or more modules are connected and assembled into a rack as shown in Fig. 3-6. This is commonly called an *array*.

Energy Storage

The laws of supply and demand come into play with solar-electric systems just as with other parts of our economy. For instance, the electric utilities that have the heaviest demand in summer due to air conditioning loads are called "summer peaking" utilities. They generally charge a premium for electricity used during this period and purchase power from other utilities to meet the demand rather than build additional plants to meet seasonal needs. If utilities could average their demand over the year, prices would be lower. Unfortunately, it is not possible to do this because electricity is difficult to store. An obvious solution where conventional electricity is available, as in most homes, would be to use solar cells to provide power when the sun is shining and buy electricity from the utility when it is not. If you generate more electricity from the cells than needed, the surplus could be fed back into the electrical grid, metered, and *sold* to the utility. This approach is covered in more detail in Chapter 4.

Although a number of possibilities exist for converting surplus electrical energy into *kinetic* (or mechanical) energy through flywheels and pumping water to elevated reservoirs, the most practical short-term solution for the storage of energy is the battery. When the sun is shining, the solar modules handle the average load for which they have been sized and, in addition, recharge the batteries. During this period, the batteries also provide *backup* power to compensate for sudden load shifts and intermittent sunshine. The batteries store power, supplying the load at night and during bad weather.

Table 3-1. Power Module Electrical Characteristics (Applied Solar Energy Corp.).

Model Part No.	Voltage (Nominal)	Output (Volts)	Output (Amps)
60-3012	6	10.0	0.3
60-3013	6	9.0	0.6
60-3014	6	9.0	1.3
60-3015	12	18.0	0.3
60-3016	12	18.0	0.6
60-3039-4	4	6.0	3.9
60-3039-6	6	9.0	2.6
60-3039-12	12	18.0	1.3
60-3053-12	12	17.0	5.2
60-3053-24	24	34.0	2.6

Many types of *storage* batteries are available for consideration. Those of the flashlight variety and other types that cannot be recharged frequently without damage are not suitable. Battery selection depends upon application and cost factors including:

- Load requirement
- Ampere-hour capacity
- Charge/discharge losses
- Operating cycle
- Environmental factors
- Reliability

Although they have the lowest initial cost, the lead-acid batteries used in many automobiles and other vehicles are not generally recommended for photovoltaic applications. It just doesn't pay to invest in more solar cells to compensate for an inefficient battery, especially where reliability is important. This becomes critical for installations that are unattended or in remote locations. Either conventional lead-antimony or lead-calcium batteries are recommended. The latter is best because it has a longer life expectancy and lower self-discharge rate (in the range of 15 percent per year compared to 200 to 300 percent for the lead-acid variety). Before you decide, it is best to contact the manufacturer for complete specifications. Some candidates are Delco Freedom models 1059 or 1071, Prestolite #2930X, and Gould #AP80. The storage capacity a battery must have is dependent upon the number of days that a load must be operated without the battery being recharged by the solar cells. Weather plays an important factor, but remember that photovoltaic arrays will provide some output even on cloudy days. The reserve capacity of batteries is specified in ampere-hours. For example, a rating of 60 means that a fully charged battery will deliver a current of 6 amps for 10 hours or an equivalent combination depending upon manufacturer's specifications.

Examples of battery selection to meet specific load requirements are provided later in the chapter. I do want to mention that for *light* loads, it may be possible to consider the use of the nickel-cadmium rechargeable batteries similar to those used in portable calculators.

Diodes and Regulators

Voltage regulation is an important part of solar-electric systems that use battery storage. Generally, the array of solar cells is sized to produce enough power during periods of bright sunshine to supply the primary load and also charge the storage batteries. There are times, however, during extended periods of good weather or if the load is less than anticipated, when the battery can be *overcharged. Voltage regulators* (also called *charge controllers*) are designed to prevent this condition. As a rule of thumb, if the rated output of the photovoltaic cells exceeds 5 percent of the battery rating in ampere-hours, a voltage regulator should be considered.

Another device with which you should be familiar is the *blocking diode.* This is a semiconductor device that functions like a one-way valve. It allows current to flow in one direction only. Note that in Fig. 3-9 the diode symbol points in the direction of flow. The blocking diode is connected in series with the solar cells and prevents the battery from discharging itself back through the array at night or other times when the cells are not producing power.

In system design, both *shunt-* (parallel or bypassing circuit) and *series*-type voltage regulators are used. The shunt types tend to be larger in size, contain finned extrusions to dissipate heat, and usually require a separate blocking diode in the circuit. In series regulation, the regulator also functions as a *switch* that opens when the batteries are fully charged and when the array voltage is lower than the battery potential. This feature eliminates the need for a blocking diode and large heat sink devices.

SOLAR ELECTRIC POWER SYSTEMS

The simplest system includes a photovoltaic cell, panel, or array connected directly to the load. The output will be in the form of direct current and, since there is no provision for storage, the system will operate during daylight hours only. In some cases, a fuse or circuit breaker might be considered if the load is easily damaged by excess power. Most applications involve upgrading such a basic system to include storage batteries and some form of voltage regulation. Also, there may be a need to supply ac output, in which case an inverter would be required. Several sources are provided in the Sources.

Once the application is determined, the next step is to calculate the number of solar panels required to match the load, with some reserve power to compensate for losses and unfavorable weather. This is called *sizing* the system, and there are three steps to this process:

1. Determine the solar energy available at the location.
2. Calculate the average daily current demand of the load.
3. Select a panel with specifications to match load requirements.

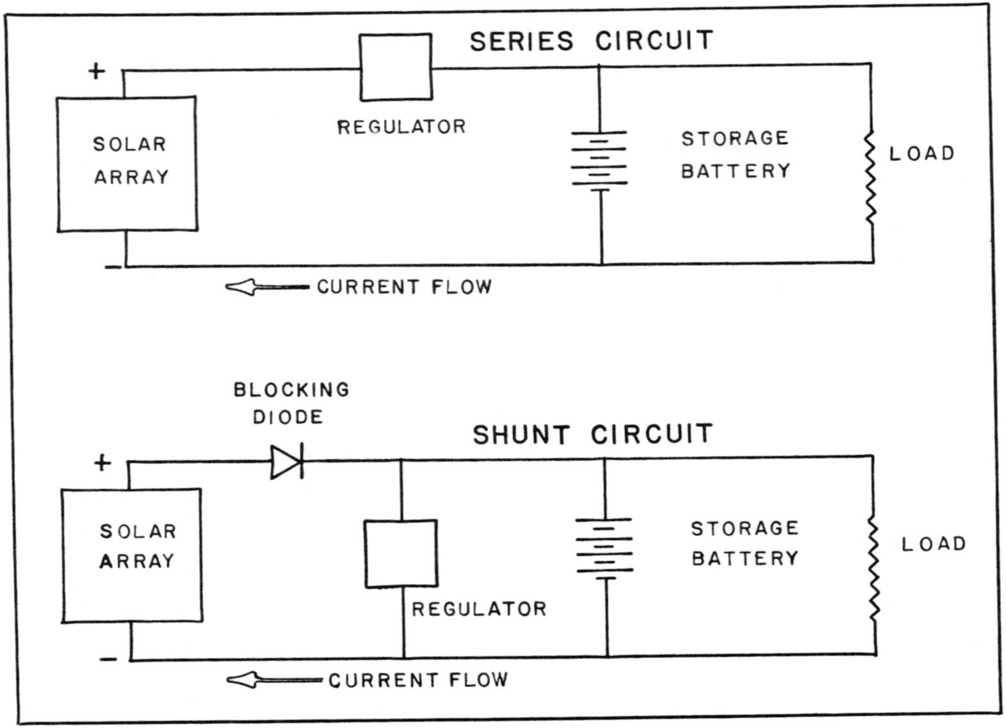

Fig. 3-9. Series and parallel connection of voltage regulation devices.

Solarex Corporation of Rockville, Maryland, has developed a simplified method of estimating the solar energy available in various geographic regions. The solar radiation maps shown in Fig. 3-10 and Fig. 3-11 express daily solar radiation (*insolation*) in terms of *equivalent sun hours*. This means one hour of sunlight at its typical maximum intensity of 1 kilowatt per square meter. For example, if a city has twelve hours of sunlight with an average intensity of 0.4 kw/m^2, the value in sun hours would be 0.4 × 12 or 4.8. These values may be read directly from the figures. If winter operation of the system is required, you should use Fig. 3-11.

Current demand of the load should be calculated in ampere-hours at the nominal operating voltage. It is recommended that you add 20 percent to this value as a safety factor to compensate for the line losses, conversion efficiency, and unpredictable weather. The ratings for many appliances may be read from the nameplate or obtained from the manufacturer.

Panel specifications are established by individual manufacturers and generally include voltage-current curves like those shown in Figs. 3-4 and 3-5 along with electrical performance and dimensions as shown in Table 3-1. Panel manufacturers generally recommend that nominal panel voltage be slightly higher than nominal battery voltage in systems using regulators and storage batteries. Systems with 12-volt batteries should use 14-volt (nominal) panels.

As an example of this sizing approach, let's assume you wish to power a small black-and-white portable television with battery backup using a solar panel similar to the setup in Fig. 3-2. The location will be Los Angeles, California. The set will be operated five hours daily and has a current drain of 1.1 amperes at 12-volt nominal operating voltage. A panel having a 1.4-ampere output at 14 volt nominal was selected:

$$\text{required system output (amperes)} = \frac{\text{average daily load in ampere-hours} + 20 \text{ percent}}{\text{equivalent sun hours at site}}$$

$$= 1.32 \text{ amperes}$$

$$\text{number of panels required} = \frac{\text{required system output (amperes)}}{\text{output (amps @ nominal voltage) of panel selected}}$$

$$= \frac{1.32}{1.4}$$

$$= 1$$

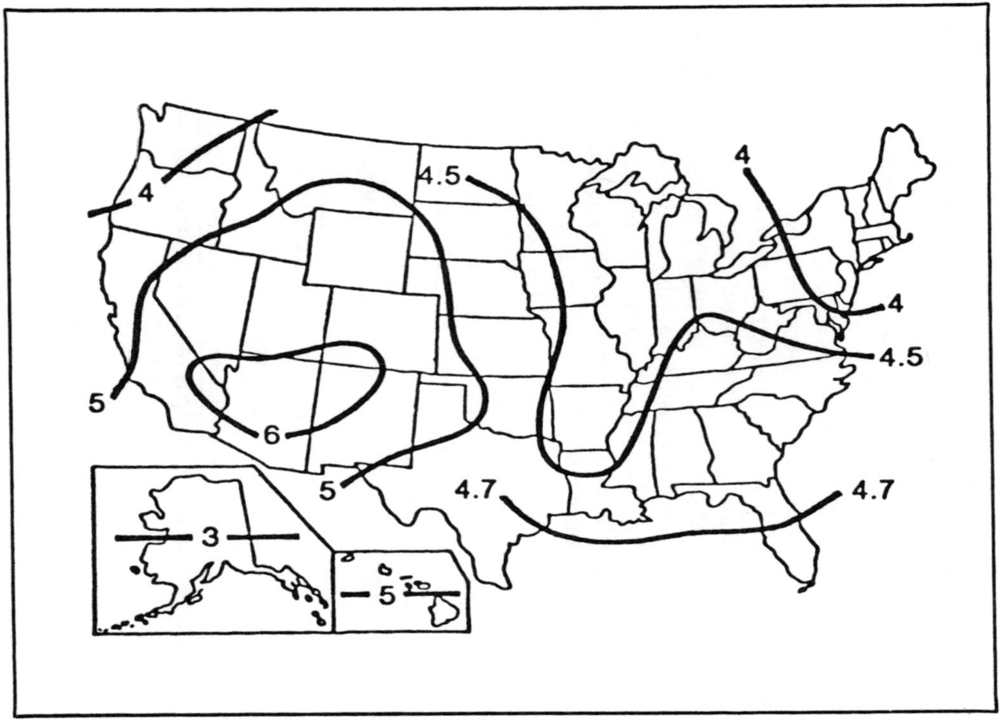

Fig. 3-10. Sun hour equivalents for various geographic regions showing the peak sun hours per day as a yearly average (courtesy, Solarex Corp.).

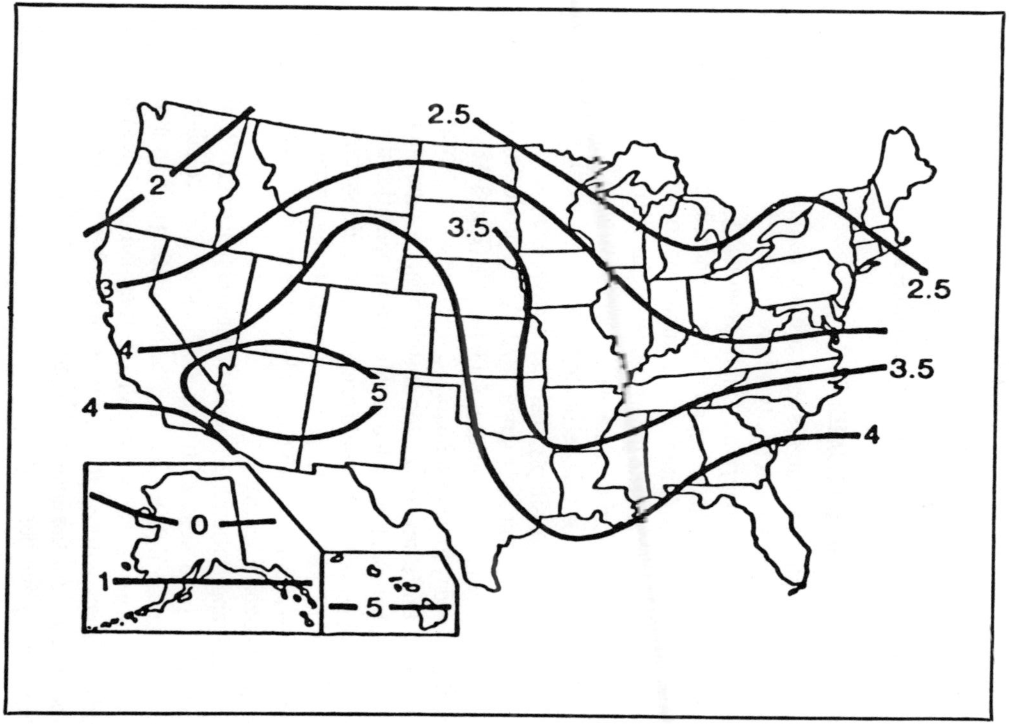

Fig. 3-11. Sun hour equivalents showing the peak sun hours per day as an average taken over the four-week period of December 7 through January 4 (courtesy, Solarex Corp.).

With a current drain of 1.1 ampere, a battery with a 60 ampere-hour rating will operate the set five hours daily for approximately nine days without being recharged by the solar panels: (60 ÷ 1.1 × 5 × 1.2).

Installation and maintenance of small solar-electric power systems is straightforward, but electrical codes and safety precautions must be observed to prevent hazardous shock or damaging shorts when connecting the components. Commercially manufactured panels are sealed against the weather, and some are even submersible without shorting out. Although thin layers of dirt, dust, or ice on the panel surface will not lower the output by much, in areas where it seldom rains, occasional cleaning with soap and water is recommended. Batteries, regulators, and the like should not be exposed to the weather. The wires connecting them to the solar panels should be kept as short as possible and well insulated.

APPLICATIONS AND ECONOMICS

To date, photovoltaic systems are not competitive with conventionally produced electricity where utilities have regular service in place. This overview limits photovoltaic applications to remote sites where it is impractical or impossible to run power lines, as in space satellites, on mountaintops, or at sea. If this is truly the case, why are the federal government and private industry spending over 250 million dollars a year for research and demonstration programs on this technology for such a small market?

The answer is simply that there have been substantial cost reductions in silicon cells since 1972. A cell producing one peak-watt of photovoltaic power cost $100 in 1972 (Fig. 3-12). It may be purchased today for approximately $9 in large volume. In addition to the cells, there are costs for installation, design, and support hardware that must be added to develop an operational system for residential power. This would bring the installed system cost into the range of $20 to $21 per peak-watt today. Another tenfold decrease to about $2 to $3 per peak-watt is needed to encourage a large residential market.

Each application has specific cost and functional "trade-offs" that must be assessed. If you don't need high reliability, low maintenance, or light weight, and the installation is in an area where commercial power is readily available, then photovoltaic systems at today's prices will find it difficult or impossible to compete. Also, since the fuel for the solar cell is sunlight, cost will be a function of its intensity and duration.

Practical Applications Today

The primary use of photovoltaic cells is in the powering of remote and/or unattended equipment. Applications include the following:

- Mountaintop communications equipment
- Navigational aids/signal devices
- Water pumping
- Forestry equipment
- Vacation homes
- Recreation vehicles
- Cathodic protection

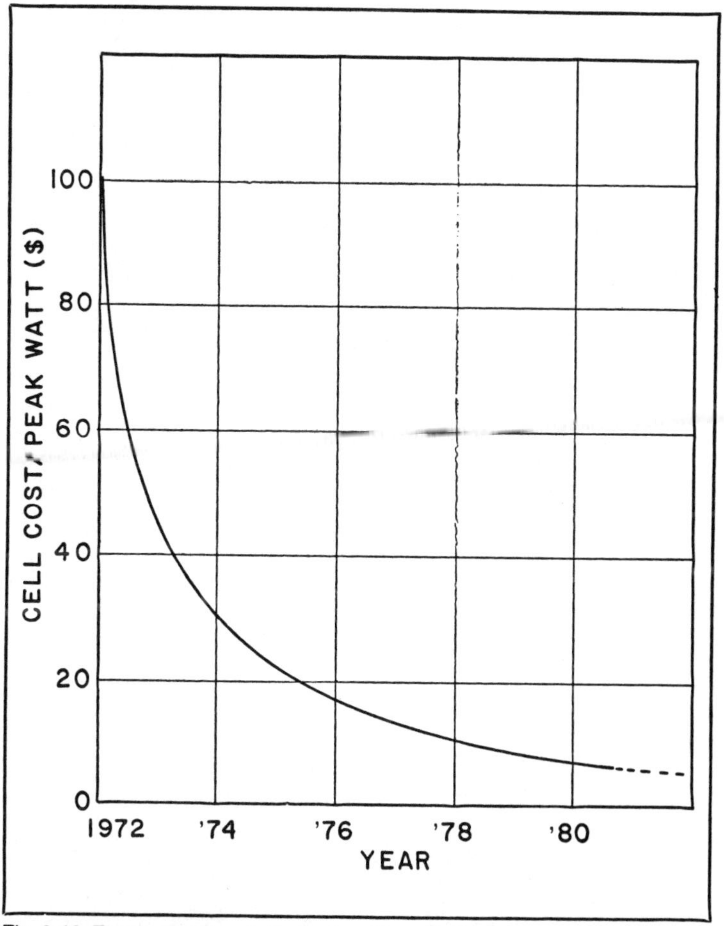

Fig. 3-12. Trends in cell cost. System cost installed is approximately double the cell cost.

Fig. 3-13. Remote site applications are good candidates for photovoltaic systems (courtesy, Motorola Solar Systems).

- Emergency power
- Environmental sensors

Most of these installations are less than one kilowatt in size.

Audio communications are a popular use of solar cells, accounting for as much as 75 percent of commercial sales. The installation pictured in Fig. 3-13 was designed by Motorola for the Arizona Public Service Co. (APS) to power a mountaintop radio repeater. Radio repeaters receive information on one frequency and generally rebroadcast on another frequency to mobile units. In this case, service personnel covering a 2,500-square-mile area receive instructions from company dispatchers for routine and emergency work. Seven solar modules manufactured by Motorola provide a total of 126 watts of peak power on a continuous basis. Battery storage is supplied by 12-volt nickel-cadmium cells with a 700 ampere-hour capacity—enough to operate ten days without sunshine.

Because remote power lines can cost from $6,000 to $10,000 per mile to install, in addition to repair and maintenance expenses, repeater sites in the past have been powered by diesel or propane thermal-electric (TEG) generators requiring frequent maintenance and refueling. This application is a natural for photovoltaic systems in areas where sufficient sunshine is available.

If you ever travel near Gila Bend, Arizona, you may pass the park facility at the base of the Painted Rock Reservoir Dam, pictured in Fig. 3-14. It is typical of many state and federal rest areas in remote locations without electrical power. The building receives power from 27 Motorola panels, providing a peak power of 648 watts to operate lights and the pumps for a waste disposal system. Lead-acid type batteries having 1,680 ampere-hour capacity are used for storage.

Water pumping is another application that holds great potential as a large photovoltaic market. This is especially true in developing countries like India and Egypt with millions of small farms that are not served by utility grids. There are also many regions in the United States that have fertile soil capable of producing good crop yields if economical irrigation were available. In addition to irrigation, photovoltaic arrays at remote sites, even mountain cabins, could be useful to power well or cistern pumps. To develop this potential, researchers are working on efficient fractional-horsepower motors and matching pumps that will combine well with the solar cells presently available.

Water Heating Project

It is now possible for you to purchase commercially available products and assemble them into a high efficiency residential water heating system that is totally independent of any utility supplied power. This application combines solar-thermal and solar-electric devices. Solar cells power the pump that circulates fluid heated by a solar-thermal collector. This home project is relatively simple if you have a working knowledge of solar water heaters. If you are not

Fig. 3-14. All electrical power to this rest station is provided by solar cells manufactured by Motorola.

Table 3-2. Specifications for Motorola 20-Watt Model MSP23A20 Solar Module (courtesy of Motorola).

Electrical

CONDITIONS AIR MASS 1.5	I_{SC} TYP	I_M MIN	V_{NO}	V_{OC} TYP	P_M MIN	P_M TYP
T_C = 28C, 100 mW/cm^2	1.3	1.1	15.8	19.5	18	20.0
T_A = 20C, T_C = 50C, 100 mW/cm^2	1.3	1.1	14.3	17.9	16.3	18.1

I_{SC} = Short Circuit Current, Adc
I_M = Current, Adc, Measured at V_{NO}
V_{NO} = Nominal Operating Voltage, Vdc, is the Reference Voltage Level at which the Modules are Designed to Provide Maximum Power Output at Specified Operating Conditions.
V_{OC} = Open Circuit Voltage, Vdc
P_M = Maximum Power, Watts
T_A = Ambient Temperature, C
T_C = Cell Temperature, C

Temperature Coefficients:
TC V_{OC} = −0.0022 V/C/Cell (Series)
TC V_{NO} = −0.00213 V/C/Cell (Series)
TC I_{SC} = 0.00036 A/C/Cell (Parallel)
TC I_M = 0.00036 A/C/Cell (Parallel)

Voltage Changes: V_{OC} decreases 2% when irradiation decreases from 100 mW/cm^2 to 80 mW/cm^2.
V_{NO} does not change appreciably with changes in irradiation level

Mechanical

Outside Dimensions (Nominal): 355 mm × 660 mm × 38.1 mm (14.0 in. × 26 in. × 1.5 in.)
Weight (Nominal): 2.2 Kg (8 lb.)
Shock: 0.4 m (15 inch) drop per MIL-STD-810B, Method 516, Procedure V
Snow Loading: 290 Kg/m^2 (60 lb/ft^2) maximum
Vibration: per MIL-STD-810B, Method 514, Procedure X

Construction

Cover Glass: 0.125 in. tempered Solatex®, 90.6% transmittance
Encapsulant: Polyvinyl Butyral
Cells: 33 in Series
Back: Aluminized Tedlar® (Polyvinyl Fluoride)
Edge Sealant: Butyl Rubber
Frame: Type 304 Stainless
Interconnect (Cell: Two Continuously Bonded-Copper Ribbons Across Top and Bottom of Cells
Connection (External): Pigtail 14 AWG

Electrical Insulation To Frame

1600 Vdc minimum

Operational Conditions

Temperature (Ambient): −40°C to 60°C
Nominal Operating Cell Temperature: 45°C at T_A = 20C, 80 mW/cm^2, Wind at 1 m/s, Module Tilted
Wind: Constant Velocity, 160 Km/hr (100 mph) maximum
Gust Velocity, 200 KmPhr (125 mph) maximum

Storage Temperature

−40°C to +60°C

familar with their principles and design, I would suggest my *Build-It Book of Solar Heating Projects* published by TAB Books.

Using a photovoltaic panel to drive the circulator is expensive but has many advantages. Since the module current is directly proportional to sunlight intensity, it powers the circulator at a variable rate, causing rapid fluid flow on a sunny day and slower circulation when the weather conditions are not as good. This can increase the efficiency of the solar water heater by as much as 17 percent over conventional methods. In addition to eliminating the differential controller, temperature sensors are not required in this system. Either potable water used in direct (open) systems or antifreeze solutions used in heat-exchanged (closed) systems can be circulated, but you must be sure the pump/motor combination selected has enough capacity to provide adequate flow. The pump controls the *rate* of water flow. This is affected by the ***head***, or pressure, it must pump against. Calculating pump size is covered in more detail in Chapter 5. If potable water is circulated, the pump should be of bronze or stainless steel construction.

I found two commercial systems for consideration. One is manufactured by Motorola and uses a 20-watt module (MSP23A20) to power a 12-volt dc pump/motor. The other is supplied by Solar Disc of Hawaii and uses an Arco 33-watt module (ASI 16-Z000) connected to an inverter to power a 12-volt ac motor magnetically coupled to a circulating pump.

Figure 3-15 shows a typical installation of the Motorola module mounted between two solar-thermal panels. A mechanical drawing of the pump/motor assembly designed for use with the module is shown in Fig. 3-16 along with the pump performance (Fig. 3-17) and system curves for various levels of sunlight intensity and ambient temperatures (Figs. 3-18 and 3-19). Power is supplied by a Motorola model MSP23A20 20-watt solar module. Specifications for this unit are shown in Fig. 3-20.

The unit is designed for domestic hot water (DHW) systems having a flow rate of three gallons per minute (gpm) or less. This will be sufficient for most systems, as typically solar-thermal panels are designed to operate at flows of 0.4 to 0.5 gpm ***per panel*** or 1.2 to 1.5 gpm for a three collector system. You must be careful, however, to insure the pump has enough "lifting power" to fill the collectors with fluid initially or, in some pumps, to refill if the fluid is allowed to drain out when the motor stops. This is called the ***static head*** and is explained more thoroughly in Chapter 5.

Installation of the PV module and pump assembly is quite

Fig. 3-15. A solar-powered pump circulates the hot water produced by two solar-thermal panels in this residential water heating application.

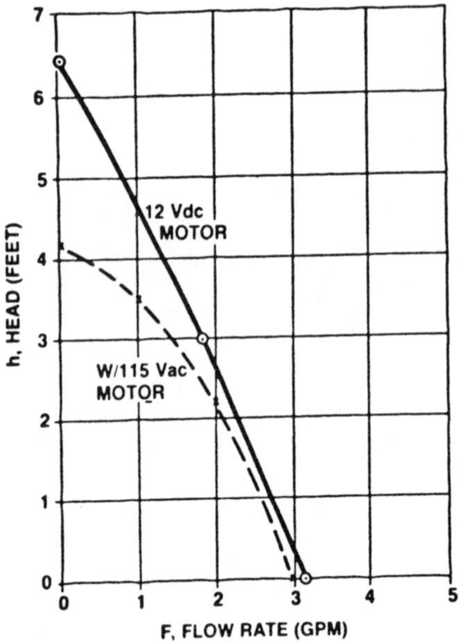

Fig. 3-16. A schematic of the pump motor of Motorola's sun-powered circulating pump model MSP23A20 (courtesy of Motorola).

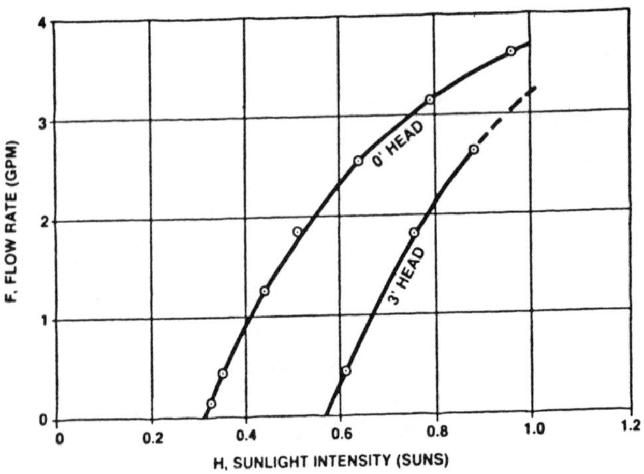

Fig. 3-17. The flow rate compared to sunlight intensity for the Motorola MSP23A20 (courtesy of Motorola).

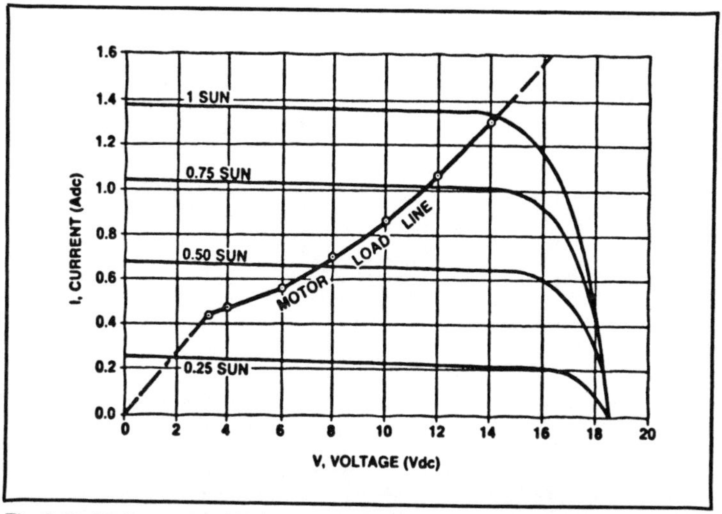

Fig. 3-18. Performance of Motorola MSP23A20 with varying amounts of sunlight when the ambient temperature is 32° F (courtesy of Motorola).

simple. You can either install the Motorola module between collectors (Fig. 3-15) or at another location. Just be sure the unit faces south, is tilted at approximately the same angle as the collectors (typically, latitude plus or minus 10°) and is *unshaded*. Avoid areas where shadows cast by vent pipes, chimneys, etc. would shade even

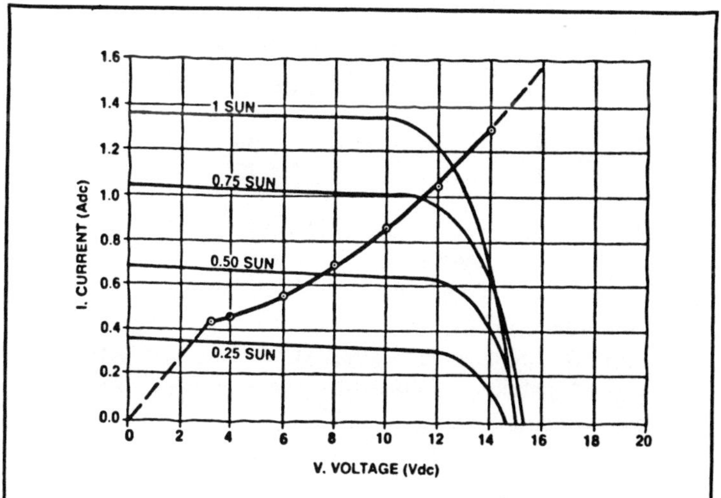

Fig. 3-19. Performance of Motorola pump MSP23A20 with varying amounts of sunlight when ambient temperature is 113° F (courtesy of Motorola).

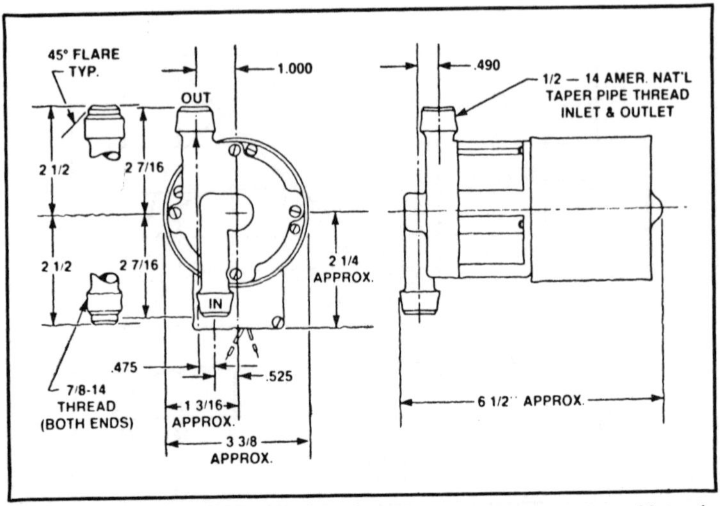

Fig. 3-20. Chart showing the head vs. flow rate at 12 Vdc of the Motorola MSP23A20 solar-powered circulating pump (courtesy of Motorola).

a portion of a single cell; for, as mentioned earlier, the entire module current is equal to that of the weakest cell in a series connection.

The pump/motor should be in an area protected from the weather and connected to the PV module using the proper wire size to minimize the voltage drop between module and motor.

wire size (AWG)	maximum distance (feet) between motor and module
16	25
14	50
12	75

This concept of a self-powered unit is particularly well suited for areas like Hawaii, where over 100 installations have been made. Power is expensive and sometimes not even available to the more remote locations, and freezing is never a problem. If you live in a freezing area or are considering this type of system for a mountain cabin, it is compatible with closed-loop antifreeze and drain-down-type solar systems as well.

4

Wind Energy

In the field of wind energy conversion systems (WECS), there are very few *new* applications to offer. Most of the ideas presented today are refinements to devices that were used centuries ago to power ships, grind grain, pump water, saw wood, and, later, to generate electricity. It is estimated that three percent of the windmills built in the United States are still in operation. The others were replaced with cheap electric power. With the current and projected cost of electrical energy, coupled with federal (and sometimes state) tax incentives, perhaps it is time once again to re-examine the potential of the wind machine.

Wind systems represent a practical energy source when specific needs correspond to the geographic, environmental, social, and legal restrictions present in the area where they are to be installed. In remote areas the barriers may be lessened by the absence of restrictive codes and zoning requirements. In cities, neighboring buildings or the visual impact of wind towers and noise or electrical interference to the neighbor's televisions may overshadow the economic advantages. This chapter presents a step-by-step analysis to help you evaluate wind devices once you understand their basic principles.

WHAT MAKES THE WIND?

The sun's rays do not contribute much heat directly to the earth's atmosphere. Most of the heat received is radiated from the land and sea bodies of the earth.

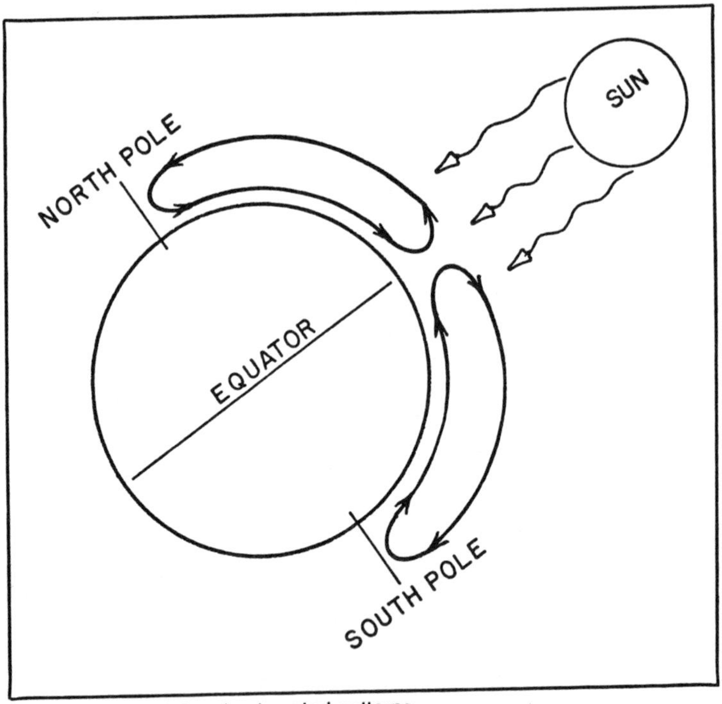

Fig. 4-1. Equatorial and polar wind patterns.

As air is heated, it becomes less dense and rises above the earth's surface. Once this happens, heavier, cold air rushes in to take its place. Basically, this air movement due to temperature differences creates *wind*. The sun is closest to the earth's surface at the equator. Because this region receives maximum solar radiation, air in the equatorial belt rises and is replaced by colder air from the North and South Poles. This creates wind currents, shown in Fig. 4-1. These fairly simple wind patterns are complicated by the fact that the earth is rotating constantly around the polar axis. Rotation causes *shear effects* at the earth's surface, diverting the colder air to the west and the warmer air at higher elevation toward the east. In the United States this creates what we commonly call the *prevailing westerlies*. Traditionally our weather moves from the west toward the east driven by the high altitude winds shown in Fig. 4-2.

We know that these westerly winds did not follow their normal path during the infamous winter of 1976-77. They were diverted toward the North Pole and picked up cold and moisture that fell as snow from the Midwest to the East Coast as far south as Florida. Tracking their path in Fig. 4-3, we can see that they basically

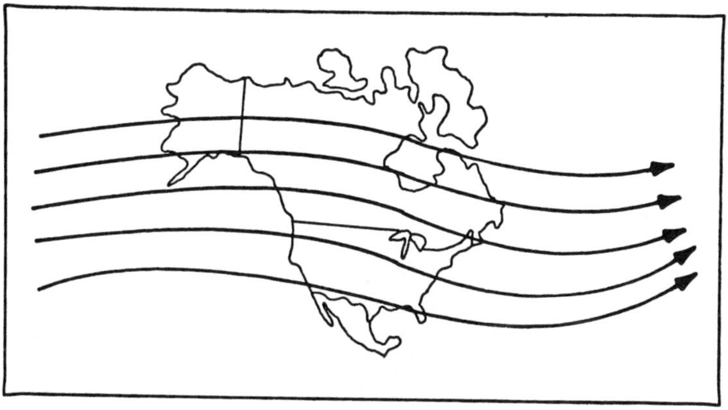

Fig. 4-2. Prevailing westerly winds in traditional pattern.

bypassed the far West, causing very little snow and a severe drought. There are different theories about why weather patterns appear to be changing, but most professionals agree that our winter weather for the last 75 years has been relatively mild and that we should, indeed, prepare for more cold in the East and less water in the West. Wind energy may offer some solutions if these predic-

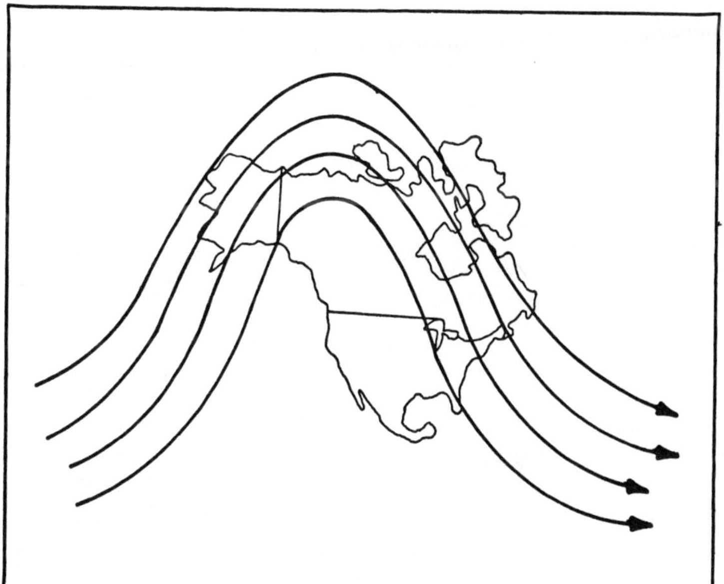

Fig. 4-3. Polar diversion of westerly winds causing severe cold in Eastern United States.

tions prove correct. Harnessing the wind to generate electricity and to pump water are just two options to consider.

Wind Zones and Measurement

It would be convenient if local wind patterns could be described as neatly as Figs. 4-1 and 4-2 depict. Wind speed and direction are affected by high- and low-pressure areas, cloud cover, large bodies of water (lakes, oceans, etc.), and terrain.

Measurement of wind speed, direction, and duration have been recorded at selected weather stations for many years, and the *Climatic Atlas of the United States* is a good source for this data. In it, you will see a section about surface *wind roses*. This is a means of diagramming the percentage of time that wind either blew from the 16 compass points or was calm. Figure 4-4 shows such a plot for 85 weather stations on an annual basis. The circle in the center of the rose shows the percentage of time the wind was calm. The length of the spokes shows graphically the percentage of time the wind blew from that given direction toward the center of the rose. Referring to the legend of Fig. 4-4 and using Great Falls, Montana, as an example, the wind comes from the southwest about 30 percent of the time and is calm only 2 percent of the time. Table 4-1 provides helpful data on the frequency of wind speed on an annual basis. Again using Great Falls, the average wind speed was 13 to 18 mph 24 percent of the time and averaged 13.9 mph throughout the year. Should you decide to investigate further, more detailed wind summaries can be obtained from the National Climatic Center, Federal Building, Asheville, NC 28801. They can supply an index that lists all sites in the United States that have developed wind summary data. If your city is not listed in Table 4-1 your local airport should be able to provide average wind speed readings for your area. If you use airport data, it is important to note that it is generally recorded at an 18-foot height and will understate the potential for wind generators on towers.

Measurements over five to ten years, obtained from 600 U.S. weather stations, were plotted and lines of constant wind speed (*isovents*) connected to develop the zone map of Fig. 4-5. Note that the north-south band through the Great Plains region extending from the Texas Panhandle to the Canadian border is the largest single high wind zone. The entire Pacific coastline and the Atlantic coastline from Main to North Carolina are considered high wind zones as well, as are portions of the Great Lakes region and southern Texas. By my definition, a high wind zone has an annual average wind speed of 14 miles per hour or greater. A low wind zone would be 12 miles per hour or less. If the average is less than 12 mph, I would not recommend investing in a wind machine.

Various techniques for measuring wind speed and direction have been used for over 300 years. Many visual methods relied upon observing a moving object from a fixed point of reference (the rate at which a sailing vessel moved along the coastline, trails of smoke from large stacks, and balloons moving aloft were some of the sightings used). One of the more unique methods was to listen for the change in pitch of vibrating objects as wind speed increased. Another wind speed indicator described the movement of trees, varying from branches being bent downwind by light winds to complete "throwing," a term to describe a tree trunk bent almost perpendicular to the ground by winds in the 20 mph range.

The best approach is to use one of the commercially available wind measuring instruments. Figure 4-6 shows an inexpensive hand-held instrument, which is small enough to fit in a shirt pocket, used at ground level. The wind speed indicator shown in Fig. 4-7 is designed for stationary installations. The plastic pickup vane, designed for pole mounting, is connected to the indicator panel by plastic piping, providing continuous readings without the need for any electrical connections. Wind provides the power. Both devices are manufactured by Dwyer Instruments. Another common instrument for measuring wind velocity is called an *anemometer*. Most anemometers have three or four cups attached to horizontal arms that turn a vertical spindle. The rate at which this spindle revolves is translated into the usable units of knots or miles per hour. The rotating spindle may drive a mechanical counter, turn a worm gear that makes a series of electrical contacts, or drive a generator whose electrical output is measured as a function of wind speed. Depending on the type of anemometer selected, the output may be either a continuous signal, such as a meter reading that changes with each gust of wind, or an intermittent signal that measures the average wind speed over a period of time.

Wind direction indicators are usually pivoted metal *vanes* or the *wind sock*. Both of these devices permit visual readings. It is also possible to obtain a remote reading from a wind vane using a *Selsyn transmitter*. The transmitter and receiver are connected electrically to the wind vane and recording instrument. The rotors on both devices are synchronized to move in unison. A remote-controlled television antenna rotor works in a similar manner where the antenna is caused to rotate to a given direction by moving the control.

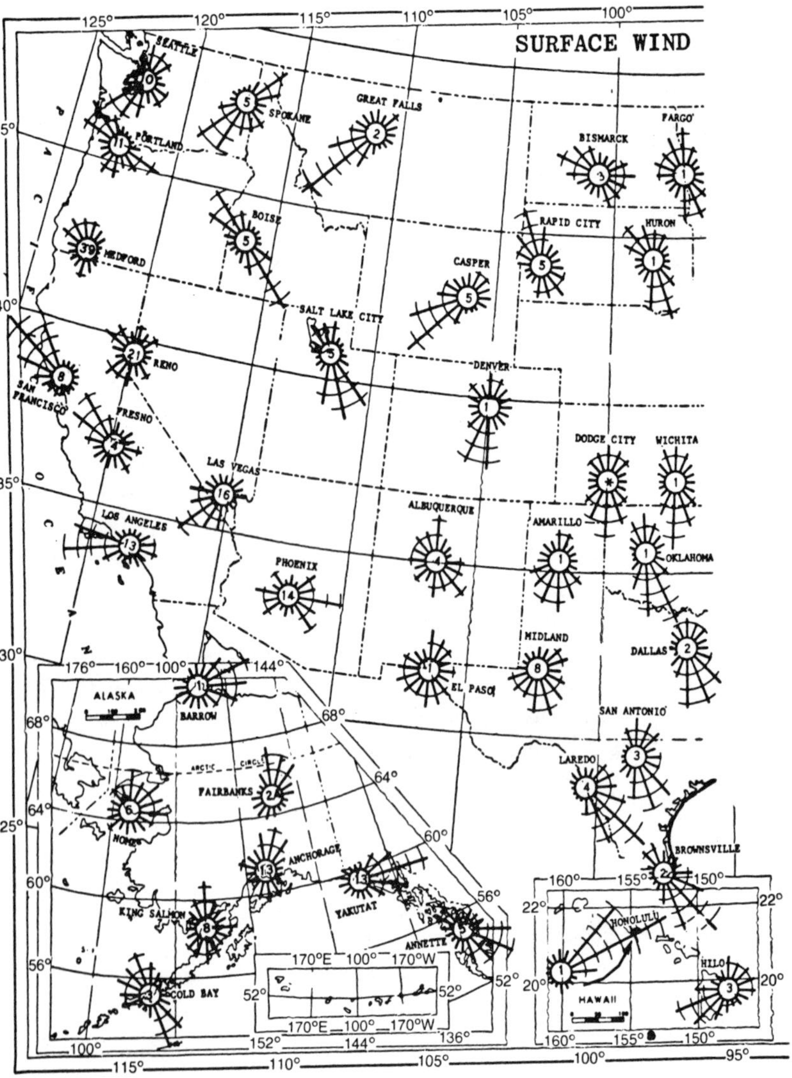

Fig. 4-4. Plot of wind speed and direction for selected U.S. locations.

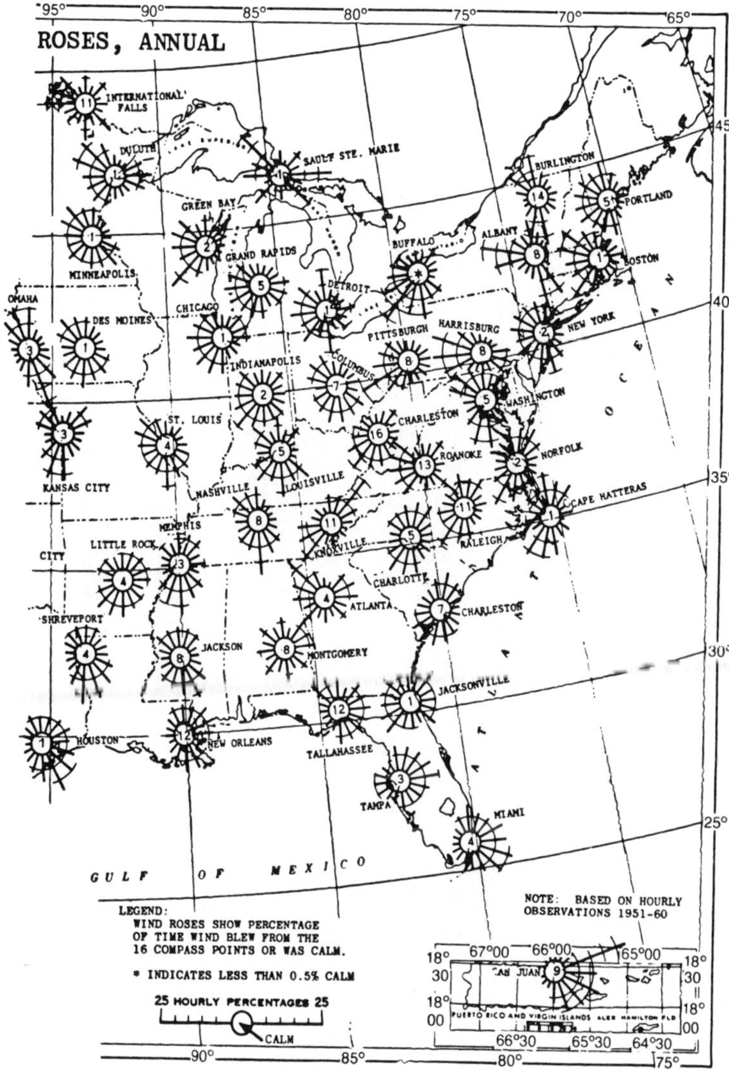
ROSES, ANNUAL
INTERNATIONAL FALLS
DULUTH
SAULT STE. MARIE
GREEN BAY
MINNEAPOLIS
GRAND RAPIDS
BUFFALO
ALBANY
BURLINGTON
PORTLAND
BOSTON
DETROIT
OMAHA
DES MOINES
CHICAGO
PITTSBURGH
HARRISBURG
NEW YORK
INDIANAPOLIS
COLUMBUS
WASHINGTON
ST. LOUIS
CHARLESTON
ROANOKE
NORFOLK
KANSAS CITY
NASHVILLE
LOUISVILLE
CAPE HATTERAS
MEMPHIS
LITTLE ROCK
KNOXVILLE
RALEIGH
CHARLOTTE
ATLANTA
CHARLESTON
SHREVEPORT
JACKSON
MONTGOMERY
JACKSONVILLE
HOUSTON
NEW ORLEANS
TALLAHASSEE
TAMPA
MIAMI
GULF OF MEXICO
NOTE: BASED ON HOURLY OBSERVATIONS 1951-60
LEGEND:
WIND ROSES SHOW PERCENTAGE OF TIME WIND BLEW FROM THE 16 COMPASS POINTS OR WAS CALM.
* INDICATES LESS THAN 0.5% CALM
25 HOURLY PERCENTAGES 25
CALM
SAN JUAN
PUERTO RICO AND VIRGIN ISLANDS

Table 4-1. Annual Percentage Frequency of Wind by Speed Groups and the Mean Speed.

State and Station	0 - 3 m.p.h.	4 - 7 m.p.h.	8 - 12 m.p.h.	13 - 18 m.p.h.	19 - 24 m.p.h.	25 - 31 m.p.h.	32 - 38 m.p.h.	39 - 46 m.p.h.	47 m.p.h. and over	Mean Speed m.p.h.
ALA, Birmingham	27	22	30	17	3	1	*	*	*	7.9
Mobile	7	28	38	20	6	1	*	*	*	10.0
Montgomery	31	29	27	12	2	*	*			6.9
ALASKA, Anchorage	28	35	25	11	2	*	*	*		6.8
Cold Bay	4	9	18	27	21	14	5	2	*	17.4
Fairbanks	40	35	19	5	1	*	*	*		5.2
King Salmon	11	20	30	24	10	4	1	*	*	11.4
ARIZ. Phoenix	38	36	20	5	1	*	*	*		5.4
Tucson	18	35	30	14	3	1	*	*		8.1
ARK. Little Rock	12	30	39	16	2	*	*	*		8.7
CALIF. Bakersfield	35	30	24	10	1	*	*			5.8
Burbank	52	26	18	4	1	*	*	*		4.5
Fresno	30	41	22	7	1	*	*			6.1
Los Angeles	28	33	27	11	1	*	*			6.8
Oakland	26	28	28	16	2	1	*	*	*	7.5
Sacramento	15	28	31	18	5	1	*	*	*	9.3
San Diego	28	38	28	6	*	*	*			6.3
San Francisco	16	21	26	22	11	3	*	*	*	10.6
COLO. Colorado Springs	9	27	38	19	6	2	*	*	*	10.0
Denver	11	27	34	22	5	2	*	*	*	10.0
CONN. Hartford	13	26	32	24	6	1	*	*	*	9.8
D.C. Washington	11	26	35	22	5	1	*	*	*	9.7
DEL. Wilmington	15	31	30	19	4	1	*	*	*	8.8
FLA. Jacksonville	10	33	35	18	3	*	*	*		8.9
Miami	14	30	34	20	2	*	*	*	*	8.8
Orlando	18	28	32	17	4	*	*	*		8.6
Tallahassee	33	36	23	7	*	*				6.1
Tampa	9	31	40	16	2	*	*	*	*	8.8
West Palm Beach	9	22	36	27	6	1	*	*		10.5
GA. Atlanta	13	24	36	21	6	1	*	*		9.7
Augusta	36	29	25	9	1	*	*			6.3
Macon	10	26	46	16	2	*	*	*	*	8.9
Savannah	12	34	37	14	3	*	*	*	*	8.4
HAWAII, Hilo	7	34	43	15	2	*	*			8.7
Honolulu	9	17	27	32	12	2	*	*	*	12.1
IDAHO, Boise	15	30	32	18	4	1	*	*		8.9
ILL. Chicago (O'Hare)	8	22	33	27	8	2	*	*	*	11.2
Chicago (Midway)	7	26	36	25	5	1	*	*	*	10.2
Moline	14	23	32	24	7	2	*	*	*	10.0
Springfield	7	22	28	27	12	3	1	*	*	12.0
IND. Evansville	19	23	32	21	5	1	*	*		9.1
Fort Wayne	9	23	33	25	8	2	*	*	*	10.9
Indianapolis	9	22	34	26	7	2	*	*	*	10.8
South Bend	7	21	35	30	7	1	*	*		10.9
IOWA, Des Moines	3	17	38	29	10	3	1	*	*	12.1
Sioux City	10	20	31	25	10	4	1	*	*	11.7

continued on page 91

Table 4-1. continued.

State and Station	0 - 3 m.p.h.	4 - 7 m.p.h.	8 - 12 m.p.h.	13 - 18 m.p.h.	19 - 24 m.p.h.	25 - 31 m.p.h.	32 - 38 m.p.h.	39 - 46 m.p.h.	47 m.p.h. and over	Mean Speed m.p.h.
KANS. Topeka	11	19	30	27	10	2	*	*	*	11.2
Wichita	4	12	30	31	16	5	1	*	*	13.7
KY. Lexington	8	25	39	22	6	1	*	*		10.1
Louisville	17	28	31	20	3	1	*	*		8.8
LA. Baton Rouge	17	29	34	17	3	*	*	*		8.3
Lake Charles	19	31	29	17	4	1	*	*	*	8.5
New Orleans	16	27	32	19	5	1	*	*	*	9.0
Shreveport	12	26	37	21	4	1	*	*	*	9.5
MAINE, Portland	10	30	33	22	4	1	*	*	*	9.6
MD. Baltimore	7	24	39	22	6	2	*	*	*	10.4
MASS. Boston	3	12	33	35	12	4	1	*	*	13.3
MICH. Detroit (City AP)	8	23	37	26	5	1	*	*	*	10.3
Flint	16	26	32	22	3	1	*	*	*	9.0
Grand Rapids	14	23	32	25	5	1	*	*	*	9.8
MINN. Duluth	6	15	33	31	11	4	1	*	*	12.6
Minneapolis	8	21	34	28	9	2	*	*	*	11.2
MISS. Jackson	33	25	26	14	2	*	*	*		7.1
MO. Kansas City	9	29	35	23	5	1	*			9.8
St. Louis	10	29	36	21	3	1	*	*	*	9.3
Springfield	4	13	34	32	13	3	1	*	*	12.9
MONT. Great Falls	7	19	24	24	15	9	3	1	*	13.9
NEBR. Omaha	12	17	20	20	11	3	*	*		11.6
NEV. Las Vegas	18	26	25	20	8	3	1	*	*	9.7
Reno	52	20	13	10	4	1	*	*	*	5.9
N.J. Newark	11	25	34	24	5	1	*	*	*	9.8
N. MEX. Albuquerque	17	36	26	13	5	2	*	*	*	8.6
N.Y. Albany	23	24	27	21	4	1	*	*		8.6
Binghamton	11	23	35	25	5	1	*	*	*	10.0
Buffalo	5	17	34	27	13	3	1	*	*	12.4
New York (Kennedy)	6	17	35	28	10	3	*	*	*	12.0
New York (La Guardia)	6	15	30	31	12	4	1	*	*	12.9
Rochester	8	22	34	25	9	2	1	*	*	11.2
Syracuse	14	27	30	23	5	1	*	*	*	9.7
N.C. Charlotte	20	32	31	14	2	*	*	*		7.9
Greensboro	20	32	31	14	2	*	*	*	*	8.0
Raleigh	18	33	34	14	2	*	*	*	*	7.7
Winston-Salem	19	22	33	21	4	1	*	*		9.0
N. DAK. Bismarck	14	20	27	24	12	3	1	*	*	11.2
Fargo	4	13	28	31	15	7	2	*	*	14.4
OHIO, Akron-Canton	7	25	35	26	5	1	*	*	*	10.4
Cincinnati	11	27	36	22	4	1	*	*	*	9.6
Cleveland	7	18	35	29	9	2	*	*	*	11.6
Columbus	26	23	29	18	4	1	*	*	*	8.2
Dayton	8	25	36	23	6	2	*	*	*	10.3
Youngstown	7	26	36	24	6	1	*	*	*	10.3
OKLA. Oklahoma City	2	11	34	34	13	6	1	*	*	14.0

Table 4-1. Annual Percentage Frequency of Wind by Speed Groups and the Mean Speed. (Continued from page 91.)

State and Station	0 - 3 m.p.h.	4 - 7 m.p.h.	8 - 12 m.p.h.	13 - 18 m.p.h.	19 - 24 m.p.h.	25 - 31 m.p.h.	32 - 38 m.p.h.	39 - 46 m.p.h.	47 m.p.h. and over	Mean Speed m.p.h.
OKLA. (Cont.) Tulsa	9	24	34	26	7	1	*	*	*	10.6
OREG. Medford	47	31	14	6	2	*	*	*	*	4.6
Portland	28	27	25	16	4	1	*	*	*	7.7
Salem	25	32	28	13	2	*	*			7.1
PA. Harrisburg	28	31	25	13	3	1	*	*		7.3
Philadelphia	11	27	35	21	5	1	*	*	*	9.6
Pittsburgh	12	26	34	22	4	1	*	*		9.4
Scranton	11	33	35	18	2	*	*	*		8.8
R.I. Providence	11	20	32	28	7	2	*	*	*	10.7
S.C. Charleston	12	28	35	19	4	1	*	*		9.2
Columbia	25	35	26	12	2	*	*			7.0
S. DAK. Huron	10	18	29	29	10	3	1	*	*	11.9
Rapid City	15	22	28	21	10	4	1	*	*	11.0
TENN. Chattanooga	39	25	24	11	1	*	*			6.1
Knoxville	29	29	25	12	4	1	*	*	*	7.5
Memphis	14	26	34	20	5	1	*	*		9.4
Nashville	27	31	25	14	2	*	*	*	*	7.2
TEX. Amarillo	5	15	32	32	12	4	1	*	*	12.9
Austin	13	25	34	23	5	1	*	*		9.7
Brownsville	10	17	25	30	14	3	*	*	*	12.3
Corpus Christi	11	16	26	33	12	2	*	*		11.9
Dallas	9	21	32	28	9	1	*	*		11.0
El Paso	10	22	32	22	9	4	1	*	*	11.3
Ft. Worth	4	14	34	34	10	3	*	*	*	12.5
Galveston	4	13	39	33	10	2	1	*	*	12.5
Houston	6	18	36	28	10	2	*	*	*	11.8
Laredo	6	15	32	34	12	1	*	*		12.3
Lubbock	4	11	33	34	13	5	1	*	*	13.6
Midland	9	22	38	26	4	1	*	*		10.1
San Antonio	18	23	32	22	4	1	*	*		9.3
Waco	3	14	36	35	10	2	*	*		12.5
Wichita Falls	5	22	41	27	5	1	*	*	*	10.5
UTAH, Salt Lake City	12	33	36	14	4	1	*	*	*	8.7
VT. Burlington	24	24	28	22	2	*	*			8.3
VA. Norfolk	14	23	30	25	6	1	*	*	*	10.2
Richmond	14	37	36	11	1	*	*	*		7.8
Roanoke	31	22	23	17	5	2	*	*		8.3
WASH. Seattle-Tacoma AP	13	16	35	26	8	2	*	*	*	10.7
Spokane	17	38	27	14	3	1	*	*		8.1
W. VA. Charleston	29	37	25	8	1	*	*			6.2
WIS. Green Bay	8	22	32	26	10	2	*	*	*	11.2
Madison	15	22	30	23	7	2	*	*	*	10.1
Milwaukee	8	17	31	30	11	3	1	*	*	12.1
WYO. Casper	8	16	27	27	13	7	2	*	*	13.3
PACIFIC, Wake Island	1	6	27	48	17	2	*	*		14.6
P. R. San Juan	15	28	27	25	4	*	*	*	*	9.1

Source: Climatography of the United States Series 82; Decennial Census of the United States Climate—Summary of Hourly Observations. 1951-60.

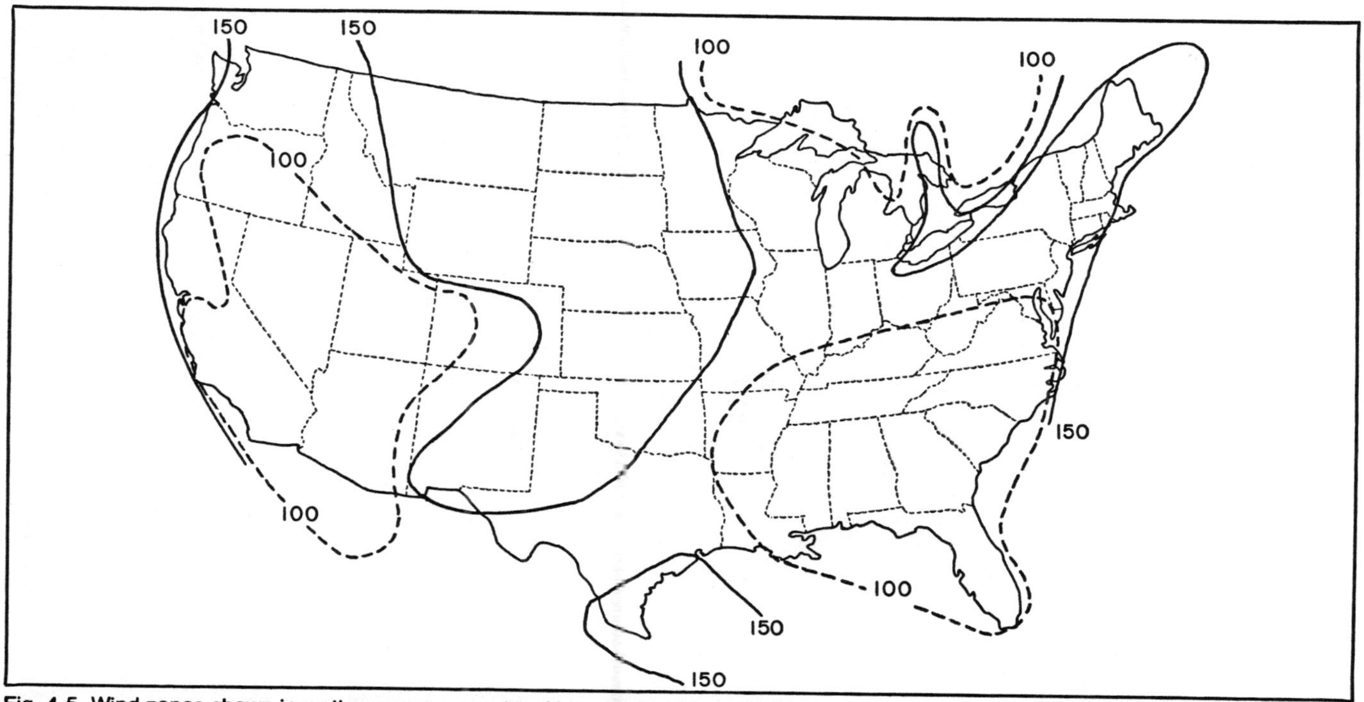

Fig. 4-5. Wind zones shown in watts per square meter: 100 equals 12.2 miles per hour; 150 equals 14.0 miles per hour.

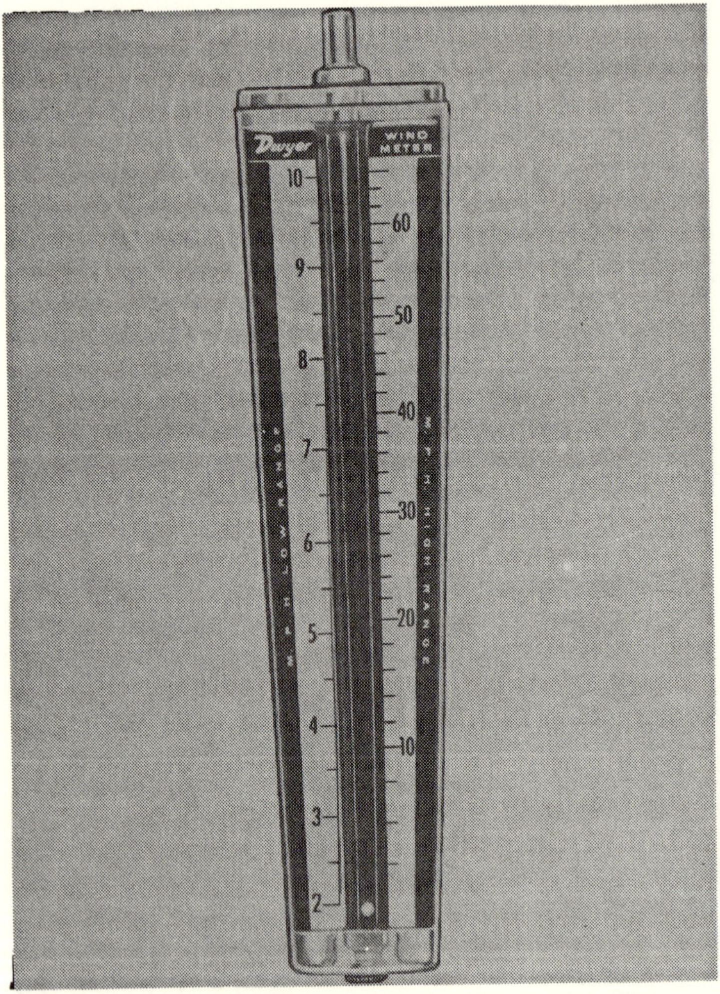

Fig. 4-6. Hand-held wind meter.

As you read more about wind measurements, you will see references to rather subjective terms like *calm, breeze,* and *gale*. Sir Francis Beaufort, an admiral and hydrographer to the British Navy in 1805, developed a scale of wind forces that is used internationally to interpret them. Table 4-2 shows the Beaufort scale and its equivalents. Note that the mean speed is given at a position 33 feet above flat ground in an open space. If you are measuring at lower positions, an approximate correction may be made by subtracting 10 percent from these values for 20-foot and 20 percent for 10-foot elevations. At a 50-foot height, 10 percent can be added.

The Effects of Terrain

In analyzing the potential power that can be extracted from wind, it is important to note flow characteristics. As you might expect, buildings, trees, hills, and valleys all disturb the smooth (or *laminar*) flow of wind and create turbulence. As you move above the earth's surface, the air flow smooths out and also increases in velocity. Over flat, open areas such as the Great Plains, the wind has little to interrupt it. But, in high density areas such as New York City, the tall buildings cause extreme turbulence. Architects generally make wind tunnel tests on proposed building designs to measure these effects.

An example of wind flow across a valley and over a hilltop is shown in Fig. 4-8. The air velocity over a hill is normally much greater than in the valley area. This phenomenon is sometimes called the *terrain enhancement effect*. The higher speed winds above the hilltop are not easily deflected and tend to concentrate or *compress* the stream, causing it to accelerate. If the terrain is steep and rocky, considerable *turbulence* causes irregular wind patterns, and the location would be less desirable for a wind machine.

As you might suspect, a flat terrain has relatively constant wind speed at any given height. If there are no buildings or trees to

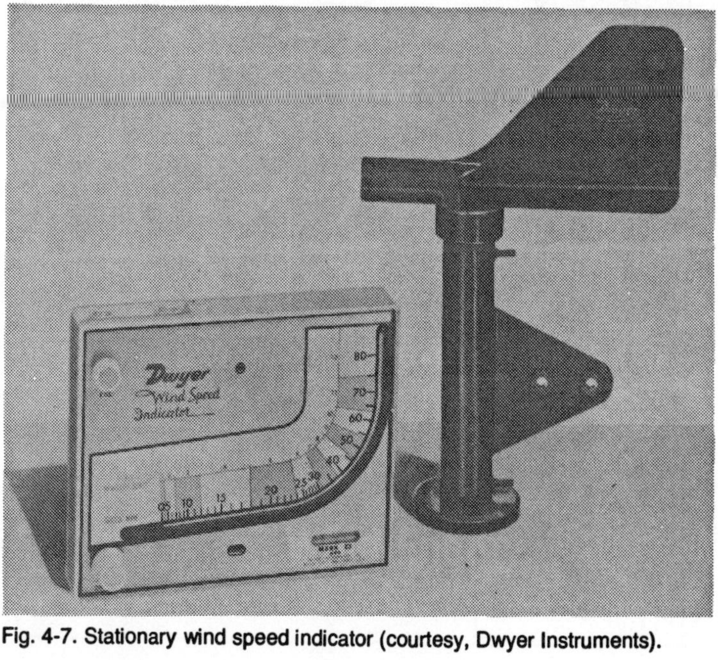

Fig. 4-7. Stationary wind speed indicator (courtesy, Dwyer Instruments).

Table 4-2. The Beaufort Scale of Wind Forces.

Beaufort Number (B)	Description of Wind	Equivalent Mean Velocity (in knots)	Limits of Mean Speed (at 33 ft. above flat ground in an open situation)	
			Knots	Statute (miles per hour)
0	Calm	0	Less than 1	Less than 1
1	Light air	2	1-3	1-3
2	Light breeze	5	4-6	4-7
3	Gentle breeze	9	7-10	8-12
4	Moderate breeze	13	11-16	13-18
5	Fresh breeze	19	17-21	19-24
6	Strong breeze	24	22-27	25-31
7	Moderate gale	30	28-33	32-38
8	Fresh gale	37	34-40	39-46
9	Strong gale	44	41-47	47-54
10	Whole gale	52	48-55	55-63
11	Storm	60	56-63	64-72
12	Hurricane	68	64-71	73-82
13	—	76	72-80	83-92
14	—	85	81-89	93-103
15	—	94	90-99	104-114
16	—	104	100-108	115-125
17	—	114	109-118	125-136

obstruct the wind, variation in speed is primarily affected by the friction of the ground. A smooth surface, such as sand, produces the least friction. Friction is increased by open grassland and even more so by tall row crops like corn. The previous correction factors for height applied to the Beaufort scale gives an approximation of how wind speed increases as you move above ground level. In many cases, there will be small buildings to contend with when selecting a location for a wind machine. These obstructions will, of course disturb the wind flow. The terms *eddy current* and *wake* are used to describe this turbulence. Turbulence can affect an area behind (downwind) the structure for a distance up to 20 times the building height. If it is necessary to site the wind machine within the 20-foot space, it should be elevated to a height greater than two times that of the building. Some additional comments on wind flows over and around objects may be found in Chapter 5 with reference to Fig. 5-3.

If you are evaluating an area that is not flat, the problem becomes more complex. Referring to Fig. 4-8, there are some important characteristics that apply to valley and hilltop or ridge locations when the terrain is smooth. At night, when the earth's surface cools, heavier air collects in valleys creating a calming effect on the wind, thereby diminishing available windpower. On

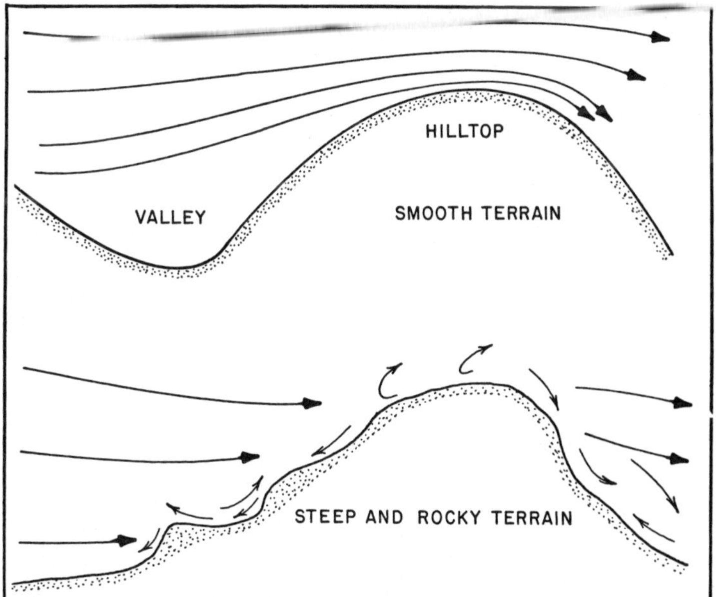

Fig. 4-8. Wind flow patterns over smooth hillside and turbulence caused by steep and rough terrain.

the ridge, not only are the winds more consistent over a 24-hour period, but they are stronger due to the compression or terrain enhancement effect discussed earlier. It is apparent that the hilltop is the preferred location to maximize windpower. Steep and rocky terrain pose so many problems that these areas should not be considered practical locations for wind machines. An excellent reference on siting is available from the National Technical Information Service (NTIS) if you decide to proceed with wind energy. Ask for publication PNL-2521 Rev. 1 entitled *A Siting Handbook for Small Wind Energy Conversion Systems* by Harry L. Wegley, et al.

TYPES OF WIND MACHINES

The vast amount of energy locked up in wind is of little value unless it can be harnessed to serve our needs. The creativity of man has spawned hundreds of specialty devices driven by the wind to convert this power. As you will observe, designs vary to suit local conditions, construction materials available, and the application at hand.

Before discussing specific designs, it will be helpful to understand the broad classifications into which wind-driven machines fall. A collector or blade is either *pushed* by the wind, causing it to move *in the direction* of the wind, or intercepts wind, causing rotation in a direction *perpendicular* to the wind. In either case, this rotating motion can be used directly to turn a generator and produce electrical power, or it can be converted into reciprocating motion for pumping water or other mechanical chores. Usually, devices whose blades or rotors move with the wind are of the *vertical axis* type, meaning that the shaft turned by the rotating blades is oriented in a vertical direction. Conversely, machines whose rotors move in a plane perpendicular to the wind direction turn a *horizontal axis* and are so classified. (There are some exceptions to these general rules.) The common farm windmill is an example of a horizontal axis machine. A modern vertical axis wind turbine is shown in Fig. 4-9.

Sail Windmill

The first sail-type windmills were vertical axis machines designed by the ancient Persians who used wind power to replace the manual and oxen-driven mills to grind grain and irrigate the land. The sail windmill used the same concept of horizontal beams rotated in a circle to turn a vertical shaft, with triangular sheets of cloth stretched like sails between the beams as the driving force—hence, the term *sail windmill*. A common application was to turn millstones, and thus we see the origin of the term *windmill.* To operate, the sails that move with the wind during one half their revolution must offer less wind resistance on their return as they attempt to move against the wind. Otherwise the shaft would not rotate. Early vertical axis sail machines used blades that folded inward on the return cycle and flipped out as they caught the wind. The cup-type anemometer described earlier is an improvement on this principle. In this case, the wind pressure on the concave cup surfaces is greater than that on the convex side, causing the device to rotate.

The Mongolian armies brought the concept from Persia into China. The early Chinese windmills were constructed in a similar fashion to the Persian ones, though eventually they took on their own character of design using sails like those of the Chinese junks for their vanes. The evolution of the windmill into Europe was probably due to the stories brought back from Asia by the crusaders. The horizontal axis mill first developed by the Europeans during the 12th century continued to be prominent throughout the continent for several centuries. These were so well built, generally of wood, stone, and iron, that they required no replacement except for sails about every 20 years.

Holland became the leader in the use of the windmill in the 17th century and held its position until the Danes took over 200 years

Fig. 4-9. Vertical axis wind turbine of the Darrieus type. Note Savonius rotors at top and bottom to start rotation.

later. In Holland, the problem was not to pump in water for irrigation, but to remove water from the low, flat lands that flooded from the tides and swollen rivers. The Dutch improved the airfoil and efficiency of the cloth sail by using wooden edges, creating new applications and a more powerful machine.

Sail windmills are still seen around the world today. Generally of the horizontal axis type, they may be found in Denmark, Holland, Great Britain, and on many of the Greek Islands. For instance, in the Valley of Lasithi, on Crete, there are presently about 10,000 windmills in use for pumping irrigation water. In the Peoples Republic of China, windmills made of wood and bamboo pump water and grind beans and rice. Simple wood and bamboo windmills with cloth sails also dot the landscape in Thailand, although recently a two-bladed propeller device has been introduced. Windmill preservation is important to these countries from an historical standpoint. In addition, improving the efficiency of the old machines with such things as modern bearings provides an energy source to countries not blessed with oil, coal, or water power.

America's Windmills

Windmills played a large part in the development of rural America, especially in the Great Plains region where they were used to pump water. Many early models were handmade by farmers with the materials that were locally available, thus providing economical irrigation for agriculture. Profiting from the lessons of earlier pioneers, towers were often used to elevate the blades and improve efficiency.

A substantial market for windmills began to develop as both railroads and the population moved westward. The application turned from milling to pumping water, permitting the towers to be less substantial and blades to be smaller in size; thus, the evolution of the *American fanwheel* windmill and the factories that produced them.

The multiblade windmills, illustrated in Fig. 4-10, with their rudder and tower (resembling an oil derrick) are what most Americans picture when the word windmill is mentioned. They are, indeed, characteristic of the early American landscape, and most western cattle ranches and farms relied on one or more to pump water for livestock.

In 1900 there were approximately 77 factories in the United States. As with any new industry, attrition took its toll, and by 1920 only 30 remained to share what was then a $10-million market.

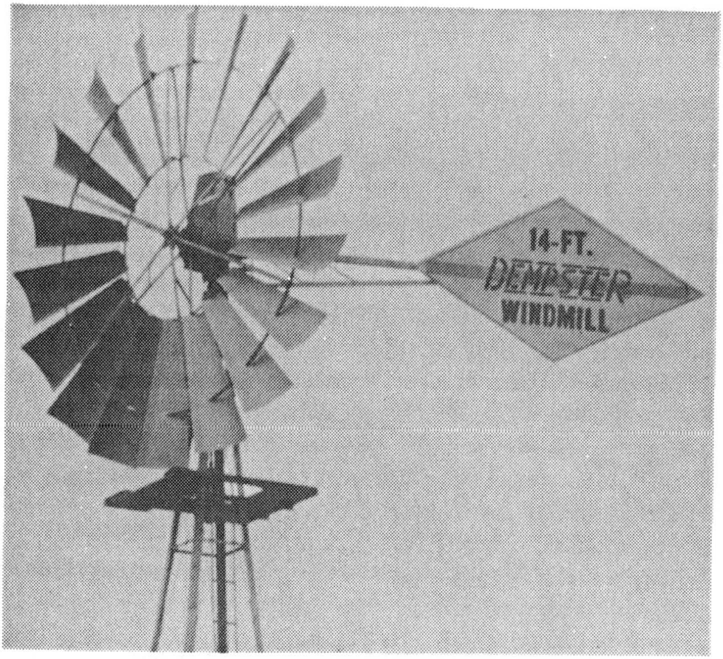

Fig. 4-10. A Dempster Industries windmill of the fanwheel type. This unit is currently available and in production.

Sears Roebuck and Co. claimed to have the most complete line of windmills on the market. In looking through a reprint of their 1908 catalog, I found two pages of windmills and towers in various sizes at prices ranging from $12.92 to $97.40 for a "regular windmill outfit with 10-foot back geared steel windmill and 60-foot no. 3 tower weighing 2430 pounds." Since the turn of the century, over six million windmills were in use in the United States. The industry met its demise during the 1930s when electricity reached into rural America. Most of the early windmills produced less than one horsepower in a 25-mile per hour wind. When electrical transmission lines were installed, small electric pumps were sold to farmers at 20 percent of the price of a comparable windmill to do the same job. The electricity cost seemed more attractive than the repair and maintenance necessary on the windmill and tower structure—especially after damage sustained in heavy windstorms. It is estimated that only 150,000 windmills are currently operating in this country, yet at least two U.S. companies, Aermotor in Arkansas, and Dempster in Nebraska, continue to manufacture multiblade fanwheels today for water pumping applications.

Propeller-Type Wind Machines

Propeller-driven wind machines are of the horizontal axis type and may have been generally considered for wind turbines used to generate electricity. The largest ever built was constructed in 1941 on a hilltop in Vermont, called Grandpa's Knob. The machine was turbine. Little was done to commercialize the concept until the period between 1970-75 when scientists at the National Research Council of Canada, and Sandia Laboratories in Albuquerque, New Mexico, began to seriously explore its potential. Figure 4-9 shows an experimental model at Sandia. Note that it has Savonius rotors at huge and, turning eight-ton blades, had a generating capacity of 1250 kilowatts. In a high windstorm, the generator threw a fractured blade over 700 feet, and the machine was not rebuilt due to excessive costs. The National Aeronautics and Space Administration (NASA) believes these massive machines have promise and has constructed a 100-kilowatt prototype (Fig. 4-11) at its test center in Sandusky, Ohio.

Researchers have demonstrated that the two-blade machines are more cost effective in high winds. Adding blades provides more torque at low wind speeds and is a preferred design for applications like pumping water—one reason for the popularity of the American fanwheel.

Regardless of the number of blades chosen, propeller type devices must be positioned directly into the wind. Orientation may be accomplished with either a tail vane, as in the American fanwheel, or with electric motors, as used in the large NASA wind generator. They must also have some provision for limiting the blade speed in high winds or no-load conditions.

Savonius Rotor

In 1925, a Finnish engineer named S. J. Savonius developed an improvement on the *S-shaped* sheet of Fig. 4-12. He reasoned that if the rotor were cut through its centerline, and the two halves displaced around a central vertical axis, wind could pass between them, reinforcing the pressure and improving the torque. This concept is shown in Fig. 4-13. You can see that as the axis is displaced, wind on part of the convex surface adds to the force of rotation. The Savonius rotor is not as efficient as a comparably sized windmill, but it does have the advantage of accepting wind from *any* direction and is simple to construct.

Fig. 4-11. A 100-kilowatt horizontal axis wind turbine developed by NASA. Man standing by front leg shows relative size.

Darrieus Rotor

An idea patented in 1927 by a French inventor, G. J.M. Darieus, offers a promising concept for a modern vertical axis wind turbine. Little was done to commercialize the concept until the period between 1970-75 when scientists at the National Research

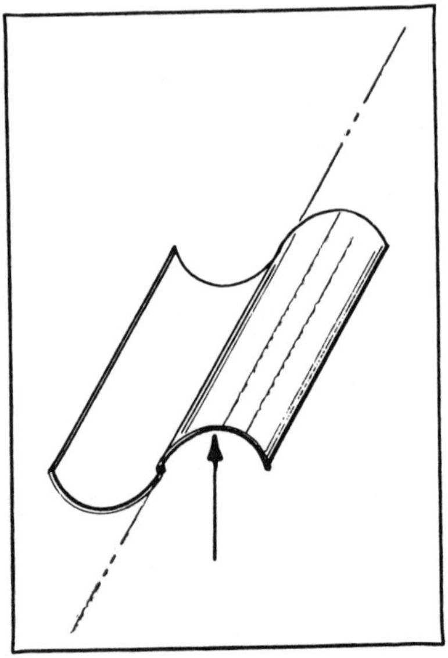

Fig. 4-12. Early design of S-shaped rotor.

Council of Canada, and Sandia Laboratories in Albuquerque, New Mexico, began to seriously explore its potential. Figure 4-9 shows an experimental model at Sandia. Note that it has Savonius rotors at the top and bottom to start rotation. Alternate designs use electric motors to start the blades turning.

The Darrieus rotor, which is both lightweight and efficient, resembles a giant eggbeater with blades having a symmetrical airfoil shape similar to a helicopter blade. The blades are curved in a shape called *troposkien*, which is Greek for "turning rope." Basically, this is the position a rope would assume if spun about a vertical axis. The rotor will accept wind from any direction, and a promising application is for small-to-medium size on-site electric generators.

WIND-DRIVEN GENERATORS

Most of the wind energy conversion systems available today are for the generation of electricity. They are relatively low output devices ranging in size from 0.01 to 40 kilowatts at specific rated speeds. Many employ direct current (dc) generators and use an inverter to change dc power into the alternating (ac) current required for most home uses. Both *remote* and *utility tie-in* models are being sold.

In remote areas where commercial power is not available, energy storage is required when a wind generator is used. As the wind is intermittent and sometimes will stop blowing for several days, wind generators are commonly used to charge batteries that store the energy until needed. They operate like a standard automobile battery but are constructed to withstand the repeated cycles of charge and discharge characteristic of wind systems. On a typical system, enough batteries are used to provide reserve power for three windless days. If the calm spell extends beyond this period, a gasoline-powered backup generator could be used to meet electrical needs. It is possible to add enough batteries to last for one week with no wind, but most people discard this option when they figure the cost. In any case, if you must have a reliable source of continuous power, an auxiliary gasoline generator is recommended.

The Public Utilities Regulatory Policy Act (PURPA) passed by Congress in 1978, provides for the interconnection or "tie-in" of small wind generators with the utility grid. This means that if the local utility is satisfied that a wind generator is safe and produces compatible output, they must agree to accept power from on-site wind machines back into the power grid when it is not needed by the owner. This, in effect, means you would buy power from the utility only when your generator could not meet demands. When output from the wind generator exceeded demand, power from the generator would be fed back to the utility lines and a meter would credit your account accordingly. Even though the homeowner must pay for the interconnection, the elimination of batteries or other auxiliary storage is a major advantage. Prices at which electricity is purchased and credited vary considerably between utilities and should be determined on a local level if you are interested in such an arrangement.

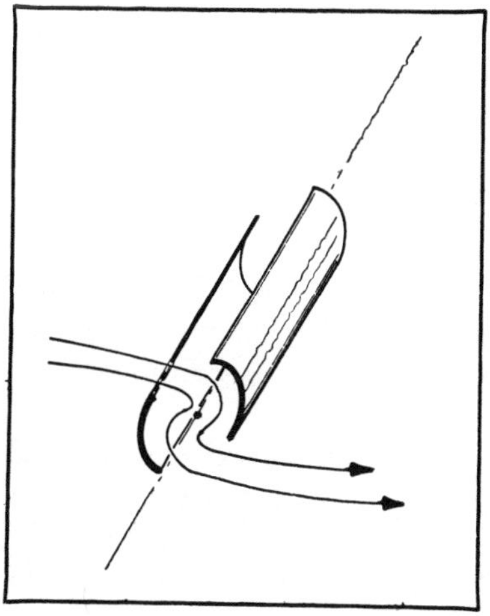

Fig. 4-13. Improvement made by S.J. Savonius in 1925 develops higher torque.

Wind Speed and Electrical Output

The earlier section on wind zones and measurement will help you determine the average annual wind speed for your area. The next step is to understand the relationship between wind speed and the electrical output you can expect from these machines.

The power that an efficient wind-driven generator can develop depends upon the wind velocity and the area swept by the rotor blades. Rotor diameter, then, is a better indicator than generator size in determining output. Power increases with the *cube* of the wind velocity. For example, if the wind velocity increased from 20 to 25 miles per hour, the power output would *double*. If the velocity increased from 10 to 16 miles per hour, the power would *quadruple*. There comes a point, however, where very strong winds create critical stresses that could damage the machine. Most wind generators, therefore, are designed with a regulator to maintain a constant rotational speed and power at some maximum level of wind velocity. Typical wind power curves for various rotor diameters are shown in Fig. 4-14.

Energy is the product of power and time commonly shown as kilowatt-hours (one kilowatt equals 1000 watts). Assuming that the generator produces an average output of 100 watts for one hour:

$$\text{kilowatts generated} = \frac{100}{1000} = \frac{1}{10}$$

$$\text{energy produced} = 1 \text{ hour} \times \frac{1}{10} \text{ kilowatt}$$

$$= 0.1 \text{ kilowatt-hour}$$

Annual energy outputs for small wind generators can be approximated from Fig. 4-15. If your average wind speed is known, draw a vertical line to intersect the curve for the rotor diameter you are considering. The estimated annual output in kilowatt-hours for this point may be read from the left-hand scale. Remember that the energy actually produced at your site can vary as much as 20 percent due to changing wind characteristics over complex terrain. Also, to get a better estimate, you should refer to the specifications each manufacturer provides on individual machines.

As an example, refer to Fig. 4-16, representing wind generators manufactured by Automatic Power under the Aerowatt name. Power generation begins at 6 to 7 miles per hour. Above 16 mph, a regulator functions to maintain a constant rotational speed

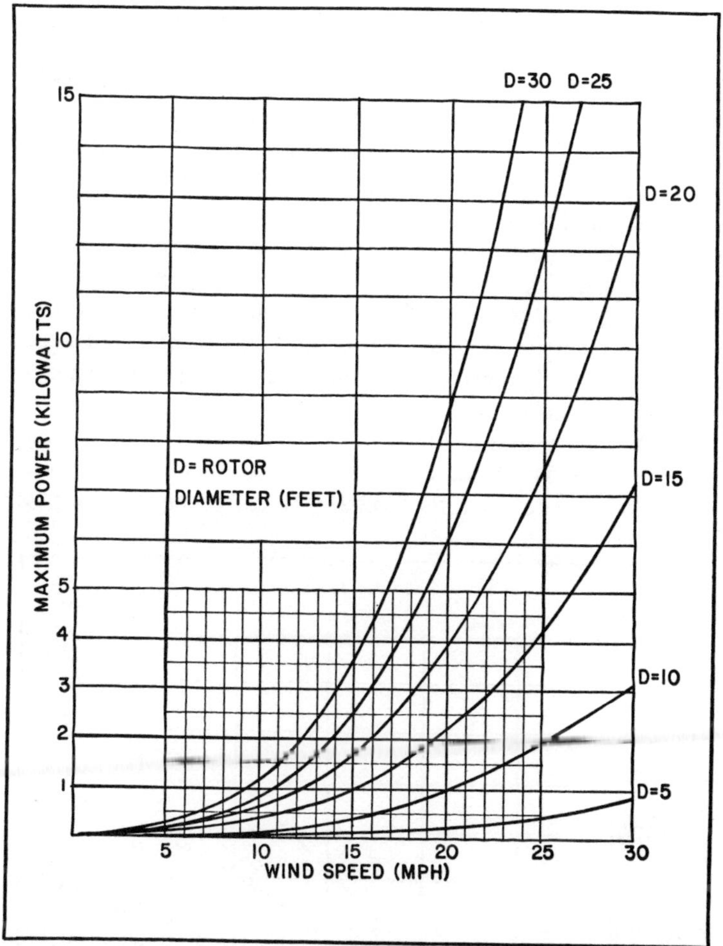

Fig. 4-14. Typical wind power curves for various rotor diameters.

and power. The instantaneous power generated will vary from zero to full output as the wind velocity varies from 6 to 16-plus miles per hour. Keep in mind that the average power output will always be *less* than the rated output due to wind speed fluctuations. Values for Aerowatt machines at the rated wind speed of 16 mph are shown in the section listing commercially available equipment found in Appendix B.

Determining Power Requirements

If you know the electrical output available from various size wind generators along with an estimate of your total energy requirements, you can decide if it is technically practical to consider windpower. To help in this evaluation, the power and energy requirements of household appliances and farm equipment in common use are listed in Table 4-3. Note that the appliance wattage (frequently shown on the nameplate) is multiplied by the hours per month you expect to operate it and divided by 1000 to arrive at a grand total of kilo-watt hours per month.

Note those appliances requiring ac power and those that can operate from dc. This is important since many windpower systems employ dc generators. You may wish to use this output directly, eliminating the need for an ac inverter, particularly in remote areas where commercial power is not available and power requirements are nominal. If this is the case, you might consider a small commercial unit as illustrated in Fig. 4-17. This machine, called a Wincharger®, is manufactured by Winco of Minneapolis, Minnesota. Using a 200-watt dc generator turned by a 6-foot propeller, charging

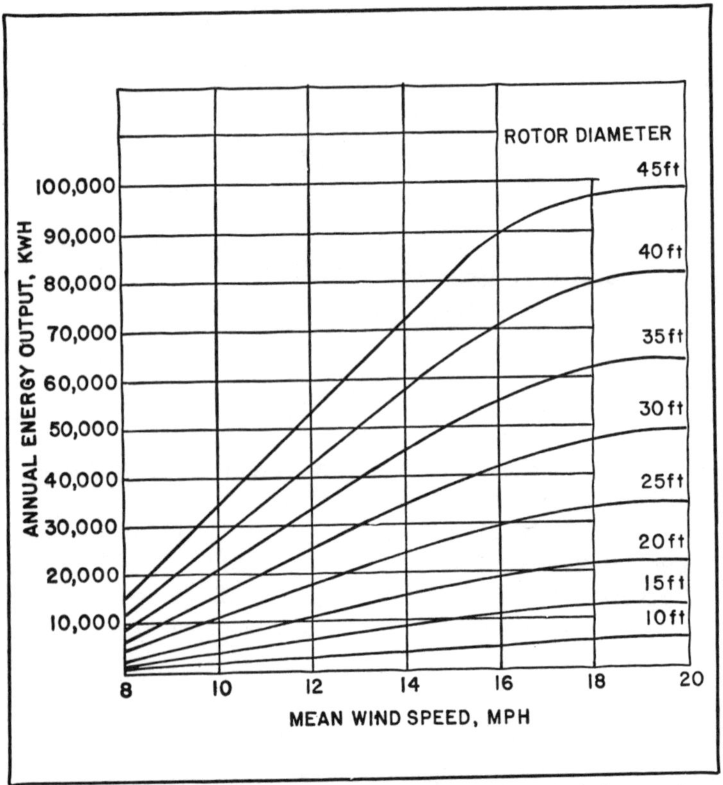

Fig. 4-15. Method of approximating annual energy output of wind generator.

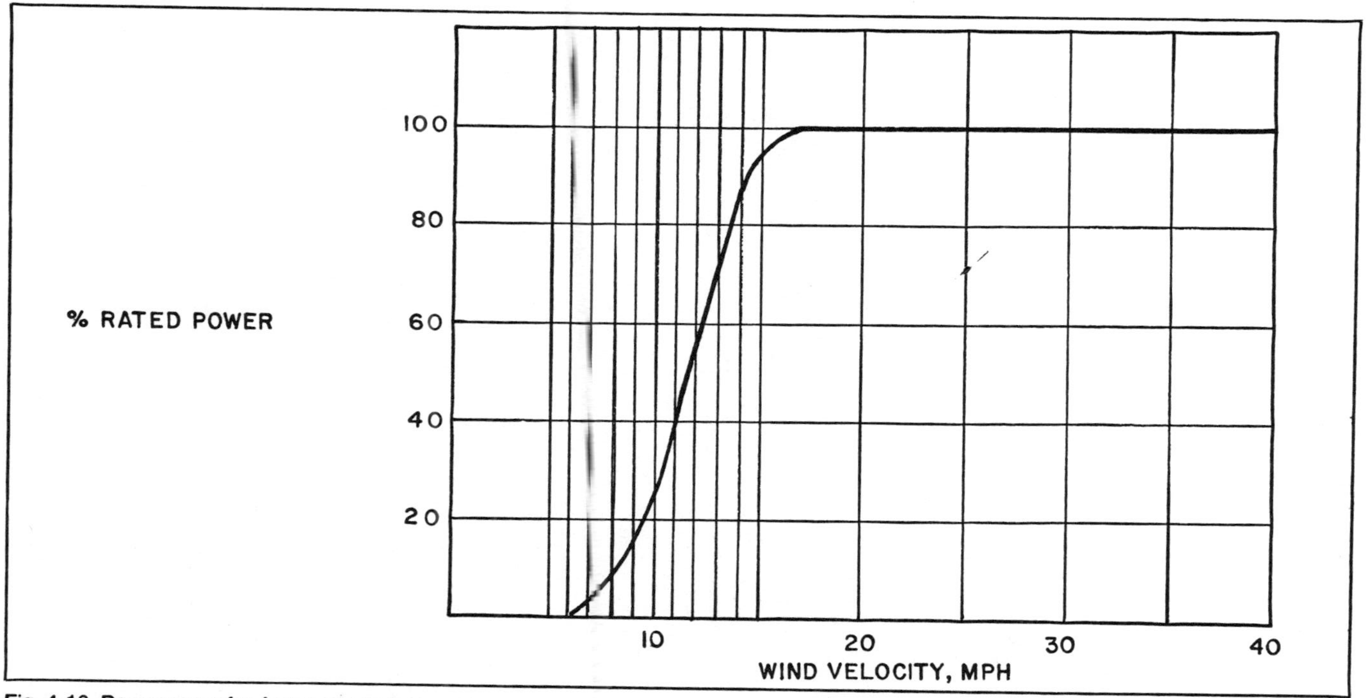

Fig. 4-16. Power curve for Aerowatt wind generators.

Table 4-3. Power and Energy Requirements of Household Appliances and Farm Equipment in Common Use.

In the House	Watts	Hrs/MO	kWHRS/MO
Air conditioning, central	-	-	620*
Air conditioning, window	1566	74	116*
Blanket	50-200		15
Blender	350	3	1
Broiler	1436	6	8.5
Clock	1-10		1.4*
Clothes dryer, electric	4600	20	92*†
Clothes dryer, gas heat	325	18	6*†
Clothes washer, automatic	512	17.3	9*
Coffee maker	800	15	12
Coffee percolator	300-600		3-10
Cooling, attic fan	1/6-¾ HP		60-90*†
Cooling, refrigerator	¾-1-1½ ton		200-500*
Corn popper	460-650		1
Dehumidifier	300-500		50*
Dishwasher	1200	30	36*
Disposal	375	2	1*
Electric baseboard heat	10,000	160	1600
Electronic oven	3000-7000		100*
Fan, attic	370	65	24*†
Fan, kitchen	250	30	8*†
Food blender	200-300		½
Food warming tray	350	20	7
Floor polisher	200-400		1
Freezer, food, 5-30 cu ft	300-800		30-125*
Freezer, frost free	440	180	57*
Fryer, cooker	1000-1500		5
Frying pan	1196	12	15
Furnace, electric control	10-30		10*
Furnace, oil burner	100-300		25-40*
Furnace, blower	500-700		25-100*†
Furnace, stoker	250-600		3-60*†
Griddle	450-1000		5
Grill	650-1300		5
Hair dryer	200-1200		½-6*
Heat lamp	125-250		2
Heater, portable	660-2000		15-30
Heating pad	25-150		1
Hi-Fi stereo			9*
Hot plate	500-1650		7-30
House heating	8000-15000		1000-2500
Humidifier	500		5-15*
Iron	1100	12	13
Knife sharpener	125		¼*
Lawnmower	1000	8	8*†
Lights, 6-room house in winter			60
Light bulb, 75	75	120	9
Light bulb, 40	40	120	4.8
Mixer	125	6	1
Movie projector	300-1000		
Oil burner	500	100	50*
Oil burner, ⅛ HP	250	64	16*

In the House	Watts	Hrs/MO	kWHRS/MO
Polisher	350	6	2
Power tools			3
Pump, water	450	44	20*t
Radio, table	40-100		5-10*
Range	8500-16000		100-150
Record player	75-100		1-5
Record player, transistor	60	50	3*
Recorder, tape	100	10	1*
Refrigerator	200-300		25-30*
Refrigerator-freezer	200	150	30*
Refrigerator-freezer/frost free	360	500	180*
Roaster	1320	30	40
Rotisserie	1400	30	42*
Sauce pan	300-1400		2-10
Sewing machine	30-100		½-2
Shaver	12		1/10
Skillet	1000-1350		5-20
Skill saw	1000	6	6
Sunlamp	400	10	4
TV, BW	200	120	24*
TV, color	350	120	42*
Toaster	1150	4	5
Typewriter	30	15	5*
Vacuum cleaner	600	10	6
Vaporizer	200-500		2-5
Waffle iron	550-1300		1-2
Water heater	1200-7000		200-300
Water pump (shallow)	½ HP		5-20*t
Water pump (deep)	1/3-1 HP		10-60*t

At The Barn	Capacity HP or Watts	Est. kWh
Barn cleaner	2-5 HP	120/yr*
Clipping	fractional	1/10 per hr
Corn, ear crushing	1-5 HP	5 per ton*
Corn, ear shelling	¼-2	1 per ton*t
Electric fence	7-10	7 per mo*t
Ensilage blowing	3-5	½ per ton
Feed grinding	1-7½	½-1½ per 100 lbs*t
Feed mixing	½-1	1 per ton*t
Grain cleaning	¼-½	1 per 100 bu*t
Grain drying	1-7½	5-7 per ton*t
Grain elevating	¼-5	4 per 1000 bu*t
Hay curing	3-7½	60 per ton*t
Hay hoisting	½-1	1/3 per ton*t
Milking, portable	¼-½	1½ per cow/mo*t
Milking, pipeline	½-3	2½ per cow/mo*t
Sheep shearing	fractional	1½ per 100 sheep
Silo unloader	2-5 HP	4-8 per ton*
Silage conveyor	1-3 HP	1-4 per ton*
Stock tank heater	200-1500 watts	varies widely

continued on page 112

Table 4-3 continued.

At The Barn	Capacity HP or Watts	Est. kWh
Yard lights	100-500 watts	10 per mo
Ventilation	1/6-1/3 HP	2-6 per day* per 20 cows
In the Milkhouse		
Milk cooling	½-5 HP	1 per 100 lbs milk*
Space heater	1000-3000	800 per year
Ventilating fan	fractional	10-25 per mo*†
Water heater	1000-5000	1 per 4 gal
For Poultry		
Automatic feeder	¼-½ HP	10-30 kWh/mo*†
Brooder	200-1000 watts	½-1½ per chick per season
Debeaker	200-500 watts	1 per 3 hrs
Egg cleaning or washing	fractional HP	1 per 2000 eggs*†
Egg cooling	1/6-1 HP	1¼ per case*
Night lighting	40-60 watts	10 per mo
Ventilating fan	50-300 watts	1-1½ per day*† per 1000 birds
Water warming	50-700 watts	varies widely
For Hogs		
Brooding	100-300 watts	35 per brooding period/litter
Ventilating fan	50-300 watts	¼-1½ per day*†
Water warming	50-1000 watts	30 per brooding period/litter
Farm Shop		
Air compressor	¼-½ HP	1 per 3 hr*
Arc welding	37½ amp	100 per year*
Battery charging	600-750 watts	2 per battery charge*
Concrete mixing	¼-2 HP	1 per cu yd*†
Drill press	1/6-1 HP	½ per hr*†
Fan, 10"	35-55 watts	1 per 20 hr*†
Grinding, emery wheel	¼-1/3 HP	1 per 3 hr*†
Heater, portable	1000-3000 watts	10 per mo
Heater, engine	100-300 watts	1 per 5 hr
Lathe, metal	¼-1 HP	1 per 3 hr
Lathe, wood	¼-1 HP	1 per 3 hr
Sawing, circular 8"-10"	1/3-½ HP	½ per hr
Sawing, jig	¼-1/3 HP	1 per 3 hr
Soldering iron	60-500 watts	1 per 5 hr
Miscellaneous		
Farm chore motors	½-5	1 per HP per hr
Insect trap	25-40 watt	1/3 per night
Irrigating	1 HP up	1 per HP per hr
Snow melting, sidewalk and steps, heating-cable imbedded in concrete	25 watts per sq ft	2.5 per 100 sq ft per hr
Soil heating, hotbed	400 watts	1 per day per season
Wood sawing	1-5 HP	2 per cord

* AC power required

† Normally AC, but convertible to DC

NOTE: Kilowatt-hour usage can be cut up to 50% through energy conservation.

Fig. 4-17. A small 200-watt machine manufactured by Winco of Minneapolis, Minnesota.

starts in a 7 mile per hour wind and reaches governing speed in a 23 mile per hour wind. At maximum, the output is sufficient to run two 100-watt light bulbs or an equivalent load. Remember that low voltage dc systems have a sizeable voltage drop and require a large wire size if the load is not close to the source. A slight drop can cause motors to run slower and lights to dim especially in 12-volt systems. If you have any distance to cover, consider a 32-volt model or use an inverter.

Utility Tie-In Models

Where commercial power is available, the popular approach is to consider a wind generator that can be connected to the utility grid with a purchase-and-buy-back-arrangement as described earlier. There are several commercial units on the market having this capability. One candidate for your consideration is a 4-kilowatt model manufactured by Hummingbird Windpower Corp. in Sweetwater, Texas. I found the review of operational and safety features on this machine provided by the manufacturer to be quite interesting and would like to share them with you.

The control supplied with the Hummingbird™ Utility Tie-In wind generating system is designed to meet requirements of the National Electric Code and local wiring codes. The unit is housed in a NEMA Type 12, UL-listed, weather tight enclosure with access doors for ease of installation and maintenance. Within the unit, automatic and manual controls protect and operate the wind generating system itself in addition to connecting and disconnecting it from the utility lines when predetermined conditions are met.

A three-phase induction motor is used to move the tail and prevent the output current level from exceeding maximum values during high wind conditions. When this occurs, the control automatically disconnects the system from the utility and initiates a two-step dynamic braking shutdown of the wind generator. Additionally, there is an overspeed protection circuit for backup in the event the normal tail control malfunctions, as well as manual controls to stop the unit. A main output circuit breaker protects against accidental grounding.

For the control to execute interconnection with the utility, the following preconditions must be met as sensed by the control:

- Utility voltage must be within plus or minus 15 percent of rated nominal.
- Utility frequency must be within plus or minus two hertz of rated.
- The motorizing detector must be reading zero.
- There can be no overload current being detected.
- The automatic brake reset circuit must have timed out 15 minutes.
- The automatic master control reset must have timed out 10 seconds.

●The manual brake switch must be in the off position.

If all of the above conditions are met, the method of interconnecting to the utility proceeds according to the following steps:

●The brake relay releases.

●The unit starts coming up to speed providing there is adequate wind (10 miles per hour or more).

●The control senses generator frequency and when the wind generator output frequency reaches a predetermined very small "frequency window," the tie-in relay closes with the series limit resistors in the circuit.

●After a delay of three seconds, the series current limit resistors are shunted out of the circuit and the unit is free to feed power into the utility system.

The control will disconnect the wind generator from utility lines whenever overload, negative current, frequency, or voltage deviations are sensed.

As the Hummingbird machine uses a direct-drive, synchronous alternator without an inverter, it has the capability of feeding power directly back into the utility's grid so that watthour meters could be run backwards.

Economic Considerations

The basic elements of a cost analysis for wind powered generators include the following:

●Equipment cost
●Installation
●Federal/state tax credits
●Repair and maintenance
●Energy cost

Both technical and economic factors must be considered.

An estimate of the expected wind generator output can be made based upon wind speed and consistency at the site selected. This information, coupled with a decision on the fraction of energy you want the machine to provide, will enable you to match up your technical requirements with commercial equipment now available. A partial list of manufacturers is provided in the Sources. Remember that in addition to the machine itself, you will probably need a tower, electrical wiring, and perhaps an inverter or storage batteries depending upon the application. If you decide to build and install the system yourself, or rebuild an old machine, plans and kits are available. But be sure you have the skills to tackle the job safely.

Table 4-4. Wind System Cost Worksheet.

System Cost		
Wind Machine		1 $______
Tower		2 $______
Wiring		3 $______
Inverter (if needed)		4 $______
Storage System (if needed)		5 $______
Other		6 $______
TOTAL SYSTEM COST (add lines 1-6)		7 $______
Adjustments		
IRS Energy Credit	8 $______ ($4,000 limit)	
State Tax Credit	9 $______	
TOTAL TAX CREDIT	10 $______	
ADJUSTED TOTAL SYSTEM COST (subtract line 10 from line 7)		11 $______
One-Time Costs		
Installation and Shipping		12 $______
Consultant (if used)		13 $______
Wind Measurements		14 $______
TOTAL ONE-TIME COSTS (add lines 12-14)		15 $______
Installed System Cost		
Add lines 11 and 15		16 $______
Fixed Charge Rate* (select one and write on line 17)		
___ .115 for residential		
___ .087 for farm or commercial		17 $______
Annual System Cost		
(multiply line 16 by line 17)		18 $______
Annual Maintenance (2.5% ** of line 7)		19 $______
TOTAL ANNUAL COST (line 18 plus line 19)		20 $______
ANNUAL ENERGY OUTPUT		21 ______kWh
Cost of Energy (line 20 divided by line 21) = ______ ×100= ______ ¢ per kWh		

NOTE: This figure is a good estimate for the first year after purchase. Since the value of money changes over time (due to inflation) and conventional energy costs can be expected to increase substantially, this figure will decrease relative to other costs during the period of wind system ownership.

*Standard adjusted annual costs for depreciation, loans, insurance, etc.
**Conservative percentage based on maintenance for similar products and current industry estimates.

To start the economic evaluation, pick out several machines with the right output, then contact the manufacturer or his dealer/distributor for delivered equipment costs as well as an installation estimate. Next, inquire about the tax credits, both state and federal, that may be available. In some cases these may be cumulative; in others, federal reimbursements must be subtracted from state credits. For your particular situation, it is best to contact the applicable state office. They should also be able to provide information on utility tie-in procedures. A list of states having wind energy incentive programs as of July, 1980, along with contact persons, is included in Appendix A.

Fixed charges and annual maintenance costs can be difficult to estimate. The best source I could find was a publication entitled "Is the Wind a Practical Source of Energy for You?" (TM-IP/80-3) prepared for the DOE Rocky Flats Wind Systems Program, dated September, 1980. The wind system costs worksheet, Table 4-4, is from this source. When completed, Table 4-4 will give you a figure for the cost of energy delivered by the wind generator. To make a thorough economic evaluation where commercial power is available, the current cost of electricity, an estimate of future costs, and the specific purchase/buy back arrangement for utility tie-in systems needs to be developed. This is another area when your state energy office may be of service. Remember that in addition to the feeling of independence that owning a wind generating system can provide, the energy saved is effectively tax-free income

5

Heating a Swimming Pool with the Sun

In the face of rapidly escalating natural gas prices and regulations that prohibit gas-fired pool heaters in some cities, the use of solar energy is one way to heat pool water and thereby extend the pool season in the spring and fall in many locations.

As with any solar approach, the first consideration is to explore ways to reduce the load before spending money on heaters. The load is determined by the size of the pool and its water temperature. Size is fixed, but temperature is a variable you can control. The most popular temperature for pool water is 80° F. Elderly people and small children generally prefer 85° F, and competitive swimmers are comfortable at 72° F. Setting back the thermostat is a matter of personal preference and generally a family decision. The load on the heater can also be reduced by energy conservation techniques like pool covers, floating discs, and wind baffles. To put the load requirements in perspective, assume you want to raise the temperature by 10 degrees in a typical 20′ × 40′ pool holding 27,000 gallons of water:

$$Q = 27{,}000 \text{ gallons} \times 8.34 \times 10 \text{ degrees}$$
$$\text{Energy required} = 2.25 \text{ million Btu}$$

(The definition of a Btu is the amount of energy required to increase the temperature of one pound of water by 1° F. Each gallon of water weighs 8.34 pounds.) Where will this energy come from and at what price? To answer these questions, we will start by reviewing the ways an outdoor swimming pool gains and loses heat.

NATURAL AND ARTIFICIAL HEATING

A pool in full sun is in fact a solar collector. Each day, approximately 75 percent of the solar energy falling on the pool surface is absorbed by the water. Typically, a pool would gain 2° F to 30° F during the day directly from the sun but will lose this heat at night, reaching an equilibrium temperature about equal to that of the outside air. Generally, a pool temperature above that of the ambient air is required for comfort, so you must devise a way to minimize heat losses and provide only enough supplementary heat to maintain comfort. Direct absorption of solar energy striking the pool surface, solar collector panels, and a conventional gas- or propane-fired heater all provide heating options. The challenge is to maximize the Btu of heat obtained for each dollar spent on these devices.

Types of Losses

Heat losses from an inground outdoor pool occur predominately at the pool surface from three primary sources:

- Sky radiation
- Convection
- Evaporation

When a pool is warmer than its surroundings, it gives up heat by *radiation.* This loss may occur 24 hours a day but is generally greater at night, when the temperature of a clear sky is substantially below that of the air near the ground. The magnitude of this loss depends upon weather conditions, increasing when the humidity is low. This is sometimes called night sky or *infrared radiation.* An effective way of reducing the loss is to cover the pool with a film that is opaque to the long wave infrared radiation, that is, one that traps the infrared rays.

Convective losses occur when air colder than the pool water blows across the surface of the pool. The air is warmed by extracting heat from the water. It rises and is replaced by a continuing supply of cold air driven by the wind. Such a process is called *forced convection.* The greater the wind speed and temperature difference between pool and air, the more rapidly heat is lost. An obvious way to reduce this loss is to install a baffle or windbreak. If you know the direction of prevailing winds, the job is easier. Baffles or fencing may also be desirable from a privacy or safety standpoint.

Loss by *evaporation* is the greatest thief of pool heat. During the swimming season, evaporative losses can easily exceed the combination of radiation and convection. To put this in perspective, with air temperature of 70° F, 50 percent humidity, and an 8-mile-per-hour wind, approximately 70 percent of the total heat loss is due to evaporation.

Loss Mechanism		**% of Total**
Sky radiation	=	10%
Convection	=	20%
Evaporation	+	70%

Detailed studies of lake evaporation were made throughout the United States by the Hydrologic Services Division of the U.S. Weather Bureau. These tests were conducted to determine the effect of evaporation on the gross water supply from a reservoir. It has been estimated that enough water evaporated from Lake Mead near Las Vegas, Nevada, to supply the needs of the Los Angeles metropolitan area. Controlling evaporation is important not only as a means of reducing heat losses, but conserving water and pool chemicals as well. Again, the use of a properly designed pool cover or chemical film is a viable method of reducing these losses. The following section provides more details for your consideration.

Pool Covers

A well-chosen pool cover is the most practical method of reducing pool heating expenses but is not universally popular. Covers are not as attractive as the sparkling water of the pool surface and someone must remove and reinstall them each time the pool is used.

For rectangularly shaped pools, electrically operated covers are available that use drive motors to roll the cover up on a window-shade-like device, but these are expensive and require maintenance. In addition they use electrical energy. I have also seen designs where the reel is operated manually and tucked neatly underneath a bench top located at the pool's edge.

For irregularly shaped pools, either plastic films cut to size or partial covers made from closed-cell plastic foam boards are possibilities. One popular product resembles a type of packing material with thousands of trapped air bubbles sealed between two plastic sheets. It floats on the surface and under normal conditions does not need anchors to keep it in place. A picture of the transparent Solar Pool Blanket manufactured by the Sealed Air Corporation® is shown in Fig. 5-1.

There are three functional types of pool covers to consider:

- Transparent/translucent films or sheets
- Opaque films/sheets
- Insulating covers

Any of the three will reduce evaporative and convection losses, but the transparent film type covers are most effective when used during the day on pools in full sun and allowed to float on the surface of the water when the pool is not in use. The cover passes sunlight easily into the pool water. And, as mentioned before, about 75 percent of the solar energy falling on the pool surface can be absorbed by the water. Opaque film covers work best when suspended just above the pool surface to reduce heat losses at night. In areas such as Arizona, with low humidity and cold, clear nights, an opaque cover constructed of insulating material like closed-cell plastic foam is especially effective in reducing heat losses due to night sky radiation.

With few exceptions, pool covers require special safety precautions to protect animals, children and nonswimmers. *Never* swim in pools which are partially covered. There have been several fatalities resulting from swimmers coming up underneath the thin plastic film that tends to cling to the face and could cause suffoca-

Fig. 5-1. Pool covers provide an economic solution to heat losses.

tion. The Federal Consumer Product Safety Commission has studied various types of covers and can provide recommendations on their use.

Solar Panels

It is not surprising that more solar panels have been sold to heat existing swimming pools than for any other applications. Pool panels cost less per square foot than the conventional solar collectors used for residential space and hot water heating because they are generally uninsulated and unglazed (not covered by glass), yet operate at high efficiency. Although a larger area for panel mounting is generally required, attractive installations can be made by framing the panels. In most cases you can use your present pool filtration pump. So, all that is needed is a south-facing roof near the pool on which to mount the collectors, the collectors, an automatic valve to divert water through the solar panels when they can contribute heat, and a controller to tell the valve when to open and close. More details on the function and economics of a solar pool heater are provided in the last section of this chapter.

Pool panels operate most efficiently by pumping water through them at a relatively high rate. Forty or fifty gallons per minute (gpm) of water may circulate through a bank of panels used in a pool heating application compared to ½ gpm for a typical solar water heater. The solar pool system heats the water about 10 to 15° F above the air temperature. The solar water heater, on the other hand, may raise the water temperature by 60° to 70° F, but the flow is so slow that it would be like trying to heat the pool with a candle!

For this reason pool panels are generally not glazed (covered) or insulated, as the temperature difference between panel and air is quite small, with little heat loss. Exceptions might be in coastal areas having cold prevailing winds or when heating is desired in the colder months using a gas heater and solar heat combined. In general, glazing a solar pool panel is impractical. The glass (or clear plastic) cover actually prevents 10 to 20 percent of the sun's energy from reaching the panel, since some of the heat is absorbed by the glass and some reflected. Reflection is a significant loss at low sun angles.

Both plastic and copper panels are used for solar pool heating, and each offers specific advantages—plastic, the advantage of lower cost; and copper, the promise of a longer service life. This is, indeed, a generalization, as there are thousands of varieties of plastic. Some will withstand the heating and cooling cycle a solar panel experiences and be resistant to the weathering effects of sunlight, and others will not. The true useful life of plastic in this application has yet to be determined, as most pool panels have been in service for less than a decade. Copper, on the other hand, has withstood the test of time as a piping material for water. Corrosion problems are minimal under most conditions, it is easy to join by soldering, and sunlight will not cause it to become brittle or crack.

Copper is, of course, a good conductor of heat, and you would expect it to exchange heat from the sun to pool water more efficiently than plastic. However, a well-designed plastic panel can compensate for lower conductivity by "wetting" a high percentage of the surface area of the panel, minimizing the distance over which heat must be transferred. By wetting, I mean maximizing the amount of water carried in relationship to the overall surface area of the panel exposed to the sun. An example of this is shown in Fig. 5-2 which compares sections of well-designed plastic panels with a conventional tube and fin copper collector. As the copper tubes are spaced several inches apart to economize on material, the thin sheet between the tubes, commonly called a fin, must collect heat and transfer it by conduction to the tube and then to the water flowing through it. In commercially available panels of this type, the tube spacing varies from 4 inches to 6 inches depending on such things as the thickness of copper used and the tube diameter. An effort is made to maximize the amount of heat collected for each pound of copper used.

Virtually all solar panels for this application are black in color where facing the sun. A blackened surface will capture more heat than one of a light color. In plastic panels, carbon black particles are generally added to the material during formulation so the color is integral. Copper panels are coated with a highly weatherable flat black paint after the metal has been thoroughly cleaned. The best quality paint resembles that used on the nose and cockpit of aircraft to reduce glare.

Wind Baffles

As discussed earlier, convective losses may rob a significant amount of heat from your pool. To make a precise calculation you need to know the wind speed, the difference between pool and air temperature, and the surface area of the pool. Such studies are beyond the scope of this book, but I felt it would be helpful to provide some fundamentals on how the wind flows over and around buildings and other objects. You can then decide if a wind baffle can

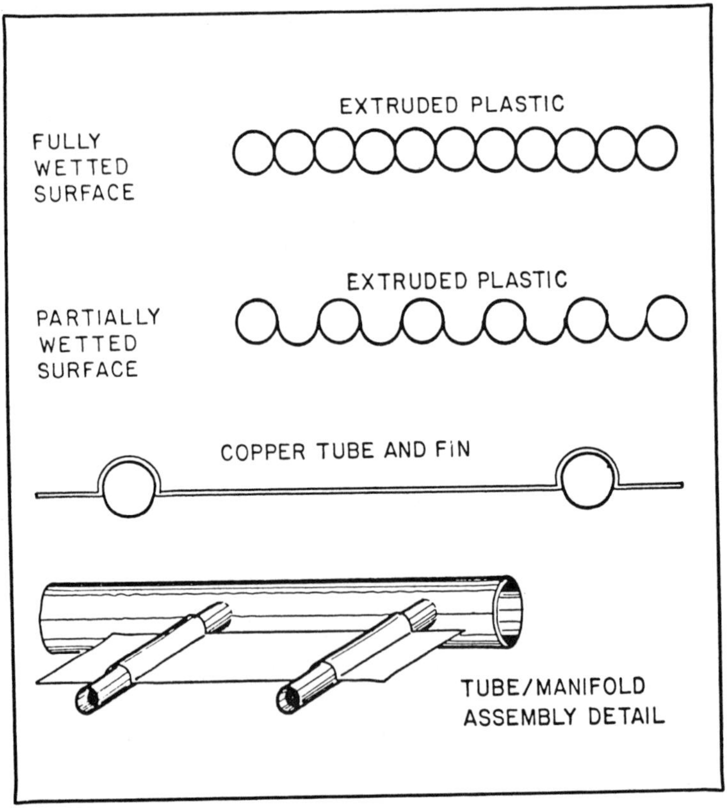

Fig. 5-2. Comparison of construction details for plastic and metal panels.

be accommodated in your individual situation. Additionally, if a new home or pool is being constructed, you can be more knowledgeable on how location might effect heating costs.

If you live in an area where the winds are relatively predictable in velocity and direction, such as sea breezes, the task of locating the windbreak or barrier is simplified. In most locations, however, wind directions are changeable, especially with the seasons.

Either a building or fence can be used to deflect the wind. When the wind flows over and around objects, *counter* or *eddy currents* develop, as shown in Fig. 5-3. They move against the main windstream with a circular motion. In any open space these eddies will occur, and it is impractical to try and eliminate them. Keeping the main wind flow from sweeping across the pool surface is the most important objective. From Fig. 5-3 you can get a rough idea how varying the height and width of the barrier protects the area behind it from the main wind. The barrier should be at least as long as the pool and located close enough to it and with enough height to provide this protection. For example, a barrier 6 feet high along the length of the pool would protect 20 feet. Increasing the height to 10 feet protects 32 feet. These are only approximations, as trees and adjacent buildings all have an influence on the wind pattern.

SOLAR POOL HEATING SYSTEMS

Complete pool heating systems are offered by a number of manufacturers throughout the United States. Most are relatively simple to install if a suitable panel mounting area is available and pipe runs are short and accessible. They are generally available through authorized dealers who sell, install, service, and warrant the system.

Many homeowners who have the knowledge can purchase panels, pipe, a controller, and valves and make the installation themselves, once a building permit has been obtained. Additionally,

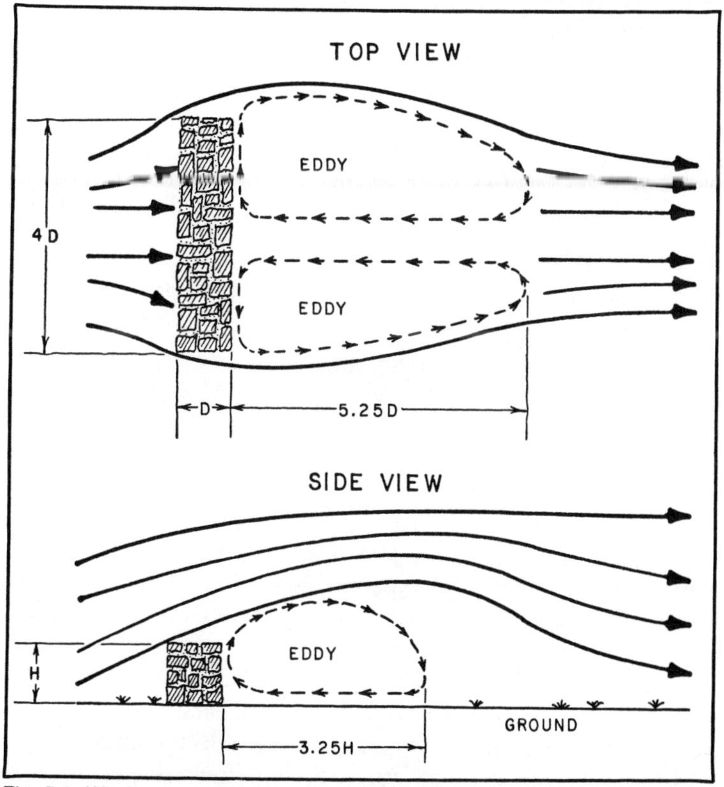

Fig. 5-3. Wind flow patterns as a function of barrier size.

some states offer tax credits for solar pool heating systems if you meet their requirements. Check the state agencies and make sure the system being considered is eligible for these financial incentives if they are offered. If you decide to do the installation yourself, I strongly recommend that you have a licensed solar contractor inspect and "sign off" the completed installation to make sure the work meets established codes and standards. Frequently, arrangements for this inspection service can be made with the dealer who sold the hardware. Some of the areas that should be checked before and after installation are the following:

- Proper grounding of the electrical system.
- Sufficient valves to isolate components.
- Freedom from fluid leaks.
- No loose wires or loose sensor attachments.
- Correct orientation and inclination of collectors.
- Proper pump capacity.
- Freedom from shade on collectors.
- Adequate size pipes.
- Balanced flow.
- Sufficient collector and plumbing line insulation.
- No air entrapment in lines or collectors.
- Collectors properly secured to structure with any roof penetrations properly sealed against leaks.

Understanding the Principles

Figure 5-4, shows how the system diverts a portion of the pool water through the solar panels whenever they can contribute heat. An electronic device called a differential controller compares temperatures and tells the diverting valve when to open and close. As pool water is recirculated and filtered once every eight to twelve hours even if solar heating is not used, placing the solar panels in a parallel circuit normally does not require an additional pump. You do, however, need to insure that with the flow specified by the panel manufacturer excessive back pressure is not created that would overload the filter pump. The types and function of solar panels used in swimming pool applications has been discussed earlier, so I will now explain the differential controller, diverted valve, and sensors.

The controller is an electronic device that monitors two low voltage sensors, one next to the solar collectors and one measuring the pool water temperature. Typically, if the collector temperature is 5° F higher than the pool water, the diverter valve is closed,

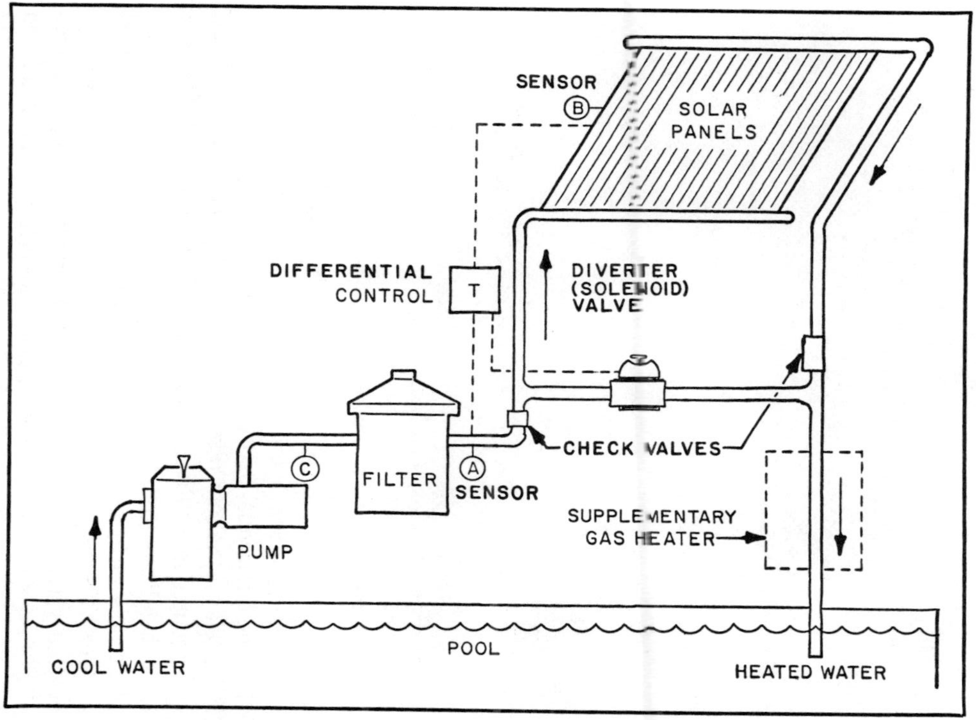

Fig. 5-4. Plumbing and electrical connections for a typical solar pool heating system using a solenoid diverter valve.

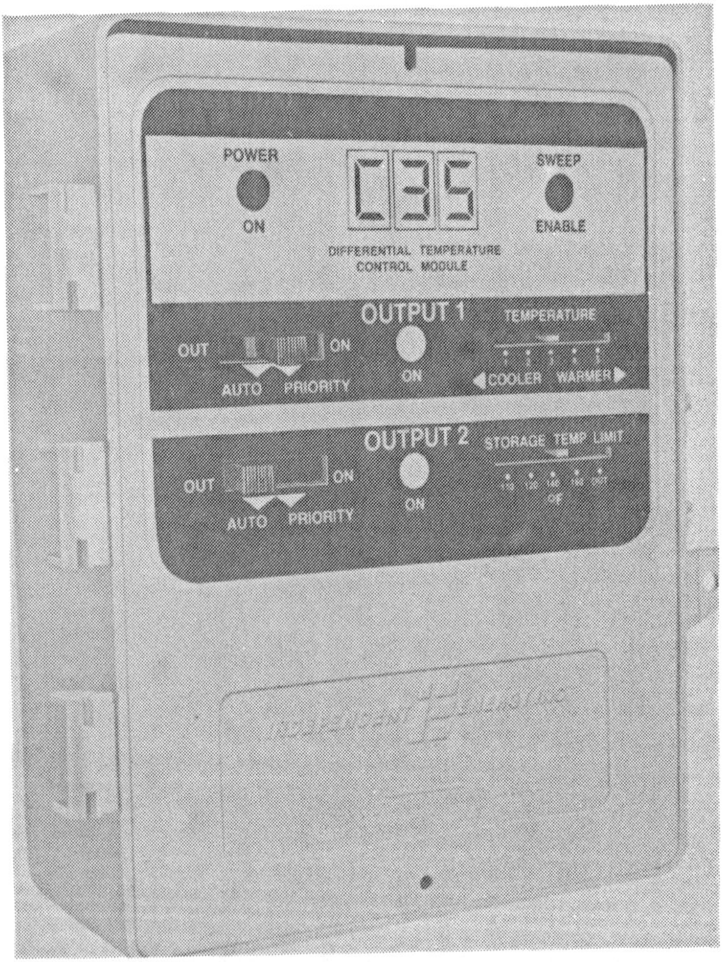

Fig. 5-5. Differential controls (courtesy, Independent Energy Inc.).

forcing pool water through the solar panels. This flow continues as long as the solar panels can contribute a significant amount of heat. When the differential between panel and pool temperatures drops 2° F, a relay in the controller box is activated that provides a 12-volt power source to the diverter valve and causes it to open. The pool water then flows from the filter directly back to the pool and the solar panels are bypassed. The differential control should be housed in a weatherproof box, be electrically matched to the sensors and diverted valve, and be approved by Underwriters Laboratories (UL). Two of the more popular models are pictured in Fig. 5-5 and Fig. 5-6.

Fig. 5-6. Differential controls for solar pool heater (courtesy, Heliotrope General).

In the system described, the diverting valve is of the solenoid type. This means it is electrically operated, using the principle of an electromagnet. When the controller provides a 12-volt charge to the valve, the solenoid coil sets up a magnetic force that either opens or closes the valve. If the valve is of the *normally open* (NO) type, it is open when *no* voltage is applied and closes with voltage. A *normally closed* (NC) valve does just the opposite. The system pictured in Fig. 5-4 uses one normally closed valve and is the most economical way to install an automatic solar pool heating system. It does have restrictions, however. There have been some cases where a siphoning action occurred, circulating water through the panels at night or under adverse weather conditions. This could actually cool rather than heat the water. It is possible to eliminate the siphon by placing an additional solenoid valve (NO type) in the collector feed line. I prefer to use a *motorized* valve.

The 12-volt (ac) motorized valve shown in Fig. 5-7 performs the same functions as two solenoid valves. It is a 3-way type that uses a small internal motor to positively divert the pool water to the collectors or bypass them completely. There is no chance for pool water to be siphoned around this valve, because the entry point from the filter pump is to the valve itself (as opposed to the solar panel inlet line being ahead of the solenoid valve). Figure 5-8 shows the system diagram using a motorized diverter valve.

Sensors used for solar pool heating are of three types:

- Pipe (pool water) sensor
- Panel simulator
- Collector sensor

Each type uses a small electronic device called a *thermistor* sealed inside a metal housing. A pipe-type sensor is used to measure the temperature of pool water after it leaves the filter. Threads are provided to screw the brass housing into a threaded pipe or fitting. If nonmetallic (plastic or rubber) solar panels are used, a sensor of the panel simulator-type measures the approximate rate that solar energy can be absorbed by the panels. It is mounted next to the panels on the same surface and, of course, must not be shaded. When copper panels are specified, a collector sensor with a copper mounting lug may be directly attached to the metal panel. As thermistors have different electrical properties, you should be sure sensors are matched electrically to the differential controller being used. It is a good idea to purchase the controller and sensors as a package supplied by the same manufacturer.

Designing Your Own System

This section is for those pool owners who have decided to purchase the hardware and to either self-install or contract the installation of a solar pool heating system. If you have the requisite skills, this section should simplify the job. If the engineering seems complex, or if your specific application is out of the ordinary, then by all means call in a professional.

Remember that the performance of heating systems that depends upon the weather is never guaranteed, as the weather itself is unpredictable. The best suggestion I can offer is to use weather data published by the U.S. Department of Commerce as a guide, as these

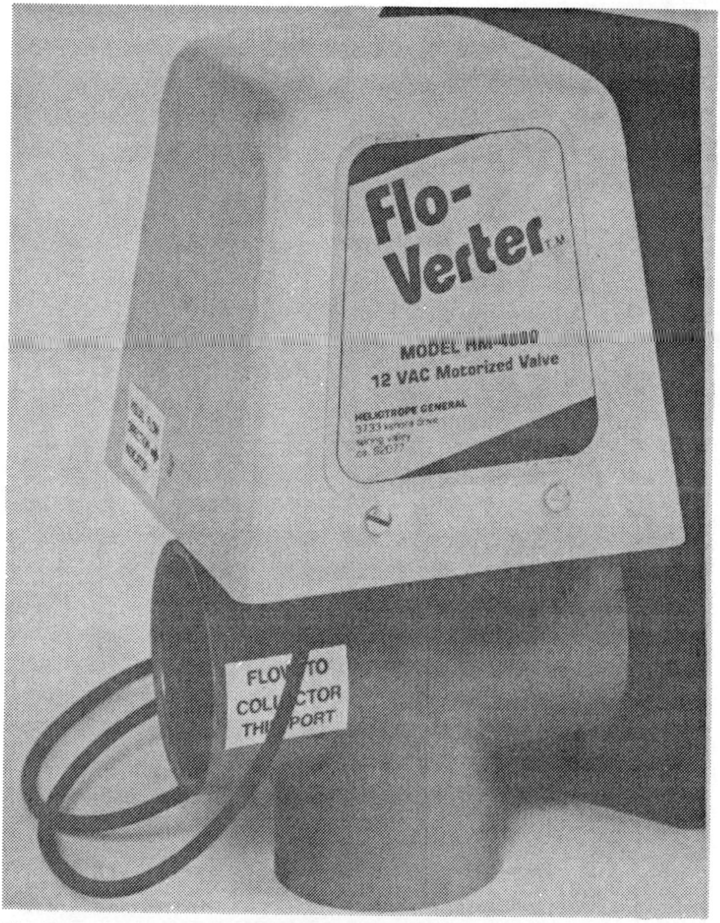

Fig. 5-7. A 12-volt motorized valve assembly (courtesy, Hellotrope General).

Table 5-1. Mean Daily Solar Radiation (Langleys).

States and Stations	Jan.	Feb.	Mar.	Apr.	May	June	July	Aug.	Sept.	Oct.	Nov.	Dec.	Annual
Alaska													
Annette	63	115	236	364	437	438	438	341	258	122	59	41	243
Barrow	≠	38	180	380	513	528	429	255	115	41	≠	≠	206
Bethel	38	108	282	444	457	454	376	252	202	115	44	22	233
Fairbanks	16	71	213	376	461	504	434	317	180	82	26	6	224
Matanuska	32	92	242	356	436	462	409	314	198	100	38	15	224
Arizona													
Page	300	382	526	618	695	707	680	596	516	402	310	243	498
Phoenix	301	409	526	638	724	739	658	613	566	449	344	281	520
Tucson	315	391	540	655	729	699	626	588	570	442	356	305	518
Arkansas													
Little Rock	188	260	353	446	523	559	556	518	439	343	244	187	385
California													
Davis	174	257	390	528	625	694	682	612	493	347	222	148	431
Fresno	184	289	427	552	647	702	682	621	510	376	250	161	450
Inyokern (China Lake)	306	412	562	683	772	819	772	729	635	467	363	300	568
La Jolla	244	302	397	457	506	487	497	464	389	320	277	221	380
Los Angeles WBAS	248	331	470	515	572	596	641	581	503	373	289	241	463
Los Angeles WBO	243	327	436	483	555	584	651	581	500	362	281	234	436
Riverside	275	367	478	541	623	680	673	618	535	407	319	270	483
Santa Maria	263	346	482	552	635	694	680	613	524	419	313	252	481
Soda Springs	223	316	374	551	615	691	760	681	510	357	248	182	459
Colorado													
Boulder	201	268	401	460	460	525	520	439	412	310	222	182	367
Grand Junction	227	324	434	546	615	708	676	595	514	373	260	212	456
Grand Lake (Granby)	212	313	423	512	552	632	600	505	476	361	234	184	417

District of Columbia													
American University	174	266	344	411	551	494	536	446	375	299	211	166	356
Silver Hill	158	231	322	398	467	510	496	440	364	278	192	141	333
	177	247	342	438	513	555	511	457	391	293	202	156	357
Florida													
Apalachicola	298	367	441	535	603	578	529	511	456	413	332	262	444
Belle Isle	297	330	412	463	483	464	488	461	400	366	313	291	397
Gainsville	267	343	427	517	579	521	488	483	418	347	300	233	410
Miami Airport	349	415	489	540	553	532	532	505	440	384	353	316	451
Tallahassee	274	311	423	499	547	521	508	542	*	*	292	230	—
Tampa	327	391	474	539	596	574	534	494	452	400	356	300	453
Georgia													
Atlanta	218	290	380	488	533	562	532	508	416	344	268	211	396
Griffin	234	295	385	522	570	577	556	522	435	368	283	201	413
Hawaii													
Honolulu	363	422	516	559	617	615	615	612	573	507	426	371	516
Mauna Loa Obs.	522	576	580	689	727	*	703	642	602	560	504	481	—
Pearl Harbor	359	400	487	529	573	566	598	567	539	466	386	343	484
Idaho													
Boise	138	236	342	485	585	636	670	576	460	301	182	124	395
Twin Falls	163	240	355	462	552	592	602	540	432	286	176	131	378
Illinois													
Chicago	96	147	227	331	424	458	473	403	313	207	120	76	273
Lemont	170	242	340	402	506	553	540	498	398	275	165	138	352
Indiana													
Indianapolis	144	213	316	396	488	543	541	490	405	293	177	132	345

continued on page 134.

Table 5-1. continued.

States and Stations	Jan.	Feb.	Mar.	Apr.	May	June	July	Aug.	Sept.	Oct.	Nov.	Dec.	Annual
Iowa													
Ames	174	253	326	403	480	541	436	460	367	274	187	143	345
Kansas													
Dodge City	255	316	418	528	568	650	642	592	493	380	285	234	447
Manhattan	192	264	345	433	527	551	531	526	410	292	227	156	371
Kentucky													
Lexington	172	263	357	480	581	628	617	563	494	357	245	174	411
Louisiana													
Lake Charles	245	306	397	481	555	591	526	511	449	402	300	250	418
New Orleans	214	259	335	412	449	443	417	416	383	357	278	198	347
Shreveport	232	292	384	446	558	557	578	528	414	354	254	205	400
Maine													
Caribou	133	231	364	400	476	470	508	448	336	212	111	107	316
Portland	152	235	352	409	514	539	561	488	383	278	157	137	350
Massachusetts													
Amherst	116	*	300	*	431	541	*	—	—	—	152	124	—
Blue Hill	153	228	319	389	469	510	502	449	354	266	162	135	328
Boston	129	194	290	350	445	483	486	411	334	235	136	115	301
Cambridge	153	235	323	400	420	476	482	464	367	253	164	124	322
East Wareham	140	218	305	385	452	508	495	436	365	258	163	140	322
Lynn	118	209	300	394	454	549	528	432	341	241	135	107	317
Michigan													
East Lansing	121	210	309	359	483	547	540	466	373	255	136	108	311
Sault Ste. Marie	130	225	356	416	523	557	573	472	322	216	105	96	333
Minnesota													
St. Cloud	168	260	368	426	496	535	557	486	366	237	146	124	348

Missouri													
Columbia (C.O.)	173	251	340	434	530	574	574	522	453	322	225	158	380
Univ. of Missouri	166	248	324	429	501	560	583	509	417	324	177	146	365
Montana													
Glasgow	154	258	385	466	568	605	645	531	410	267	154	116	388
Great Falls	140	232	366	434	528	583	639	532	407	264	154	112	366
Summit	122	162	268	414	462	493	560	510	354	216	102	76	312
Nebraska													
Lincoln	188	259	350	416	494	544	568	484	396	296	199	159	363
North Omaha	193	299	365	463	516	546	568	519	410	298	204	170	379
Nevada													
Ely	236	339	468	563	625	712	647	618	518	394	289	218	469
Las Vegas	277	384	519	621	702	748	675	627	551	429	318	258	509
New Jersey													
Seabrook	157	227	318	403	482	527	509	455	385	278	192	140	339
New Hampshire													
Mt. Washington	117	213	238	*	*	*	—	—	*	*	*	96	—
New Mexico													
Albuquerque	303	386	511	618	686	726	683	626	554	438	334	276	512
New York													
Ithaca	116	194	272	334	440	501	515	453	346	231	120	96	302
N.Y. Central Park	130	199	290	369	432	470	459	389	331	242	147	115	298
Sayville	160	249	335	415	494	565	543	462	385	289	186	142	352
Schenectady	130	200	273	338	413	448	441	397	299	218	128	104	282
Upton	155	232	339	428	502	573	543	475	391	293	182	146	355
North Carolina													
Greensboro	200	273	354	469	531	564	544	485	406	322	243	197	383

continued on page 136

Table 5-1. continued.

States and Stations	Jan.	Feb.	Mar.	Apr.	May	June	July	Aug.	Sept.	Oct.	Nov.	Dec.	Annual
Hatteras	238	317	426	569	635	652	625	562	471	358	282	214	443
Raleigh	235	302	*	466	494	564	535	476	379	307	235	199	—
North Dakota													
Bismark	157	250	356	447	550	590	617	516	390	272	161	124	369
Ohio													
Cleveland	125	183	303	286	502	562	562	494	278	289	141	115	335
Columbus	128	200	297	391	471	562	542	477	422	286	176	129	340
Put-in-Bay	126	204	302	386	468	544	561	487	382	275	144	109	332
Oklahoma													
Oklahoma City	251	319	409	494	536	615	610	593	487	377	291	240	436
Stillwater	205	289	390	454	504	600	596	545	455	354	269	209	405
Oregon													
Astoria	90	162	270	375	492	469	539	561	354	209	111	79	301
Corvallis	89	*	287	406	517	570	676	558	397	235	144	80	—
Medford	116	215	336	482	592	652	698	605	447	279	149	93	389
Pennsylvania													
Pittsburgh	94	169	216	317	429	491	497	409	339	207	118	77	280
State College	133	201	295	380	456	518	511	444	358	256	149	118	318
Rhode Island													
Newport	155	232	334	405	477	527	513	455	377	271	176	139	338
South Carolina													
Charleston	252	314	388	512	551	564	520	501	404	338	286	225	404
South Dakota													
Rapid City	183	277	400	482	532	585	590	541	435	315	204	158	392

Tennessee													
Nashville	149	228	322	432	503	551	530	473	403	308	208	150	355
Oak Ridge	161	239	331	450	518	551	526	478	416	318	213	163	364
Texas													
Brownsville	297	341	402	456	564	610	627	568	475	411	296	263	442
El Paso	333	430	547	654	714	729	666	640	576	460	372	313	536
Ft. Worth	250	320	427	488	562	651	613	593	503	403	306	245	445
Midland	283	358	476	550	611	617	608	574	522	396	325	275	466
San Antonio	279	347	417	445	541	612	639	585	493	398	295	256	442
Utah													
Flaming Gorge	238	298	443	522	565	650	599	538	425	352	262	215	426
Salt Lake City	163	256	354	479	570	621	620	551	446	316	204	146	394
Virginia													
Mt. Weather	172	274	338	414	508	525	510	430	375	281	202	168	350
Washington													
North Head	*	167	257	432	509	487	486	436	321	205	122	77	—
Friday Harbor	87	157	274	418	514	578	586	507	351	194	102	75	320
Prosser	117	222	351	521	616	680	707	604	458	274	136	100	399
Pullman	121	205	304	462	558	653	699	562	410	245	146	96	372
Univ. of Washington	67	126	245	364	445	461	496	435	299	170	93	59	272
Seattle-Tacoma	75	139	265	403	503	511	566	452	324	188	104	64	300
Spokane	119	204	321	474	563	596	665	556	404	225	131	75	361
Wisconsin													
Madison	148	220	313	394	466	514	531	452	348	241	145	115	324
Wyoming													
Lander	226	324	452	548	587	678	651	586	472	354	239	196	443
Laramie	216	295	424	508	554	643	606	536	438	324	229	186	408

continued on page 138.

Table 5-1. continued.

States and Stations	Jan.	Feb.	Mar.	Apr.	May	June	July	Aug.	Sept.	Oct.	Nov.	Dec.	Annual
Island Stations													
Canton Island	588	626	634	604	561	549	550	597	640	651	600	572	597
San Juan, Puerto Rico	404	481	580	622	519	536	639	549	531	460	411	411	512
Swan Island	442	496	615	646	625	544	588	591	535	457	394	382	526
Wake Island	438	518	577	627	642	656	629	623	587	525	482	421	560

* insufficient data collected
— incomplete monthly data
≠ Barrow is in darkness during the winter months

Langley is the unit used to denote one gram calorie per square centimeter.

Source: Climatic Atlas of the United States
U.S. Department of Commerce

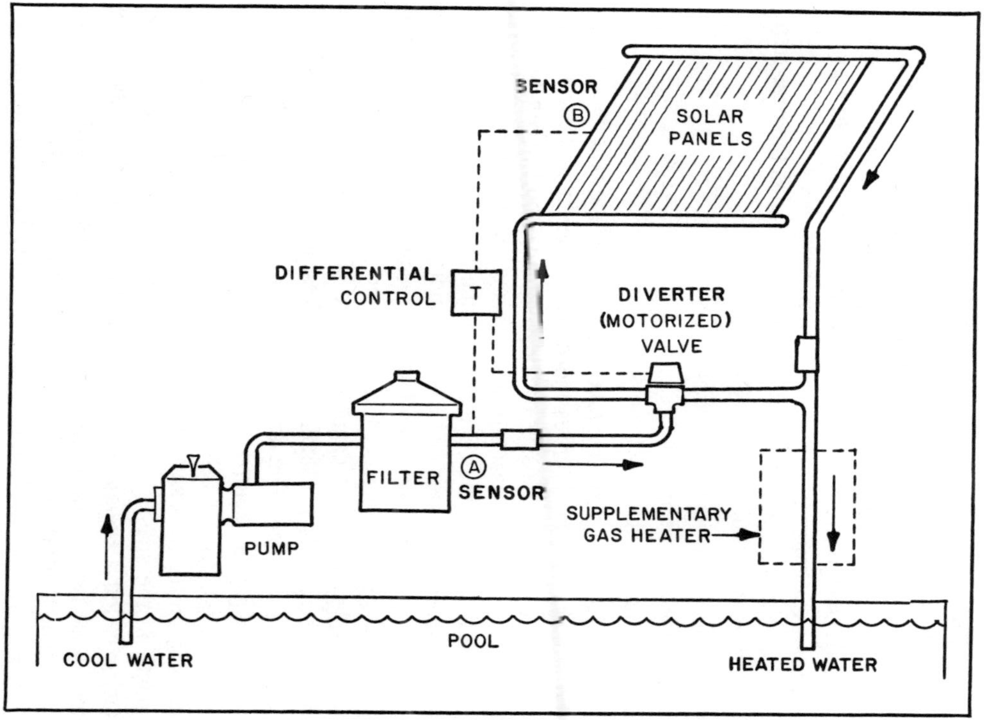

Fig. 5-8. System diagram using a motorized diverter valve.

statistics have been collected over a long period of time in more than one hundred key cities throughout the United States. Table 5-1 represents some of the data you will need.

Before proceeding with the design review, it is important to understand the limitations of the solar heater. It is expected to raise the pool temperature approximately 10° F over the ambient temperature. The objective is to do this in steps, over a period of one week, as early in the year as local weather permits. You will need continuous sunny weather for this to happen. Otherwise, you must rely on a gas-fired heater to boost the temperature or wait longer. In using the following sizing method, it is assumed that the pool and solar panel surfaces are not shaded from available sun from 8:00 AM to 4:00 PM and that you do not start or end your swimming season during months where air temperatures fall below 60° F during these hours. Calculating the effects of shade is a complex problem. For a more thorough analysis, I would refer you to the technical report "How to Design and Build a Solar Swimming Pool Heater," by the Copper Development Association, New York, New York.

Calculating the area of solar panels required is the first step. To do this, you will need to establish:

- The desired swimming season
- Surface area of the pool
- Slope and orientation of the solar panels

It is assumed that you have taken steps to reduce heat losses from the pool by using wind baffles and/or a transparent pool cover. With reasonable care, pool losses can be assumed to be 100 Btu/hr – sq ft of pool surface in most cities. An exception, which may surprise you, are the desert cities like Phoenix and Tucson, which have very high evaporative losses due to low humidity. Night sky radiation is also high in these areas, so I would recommend a figure of 130 Btu/hr – sq. ft. for desert locations. Another assumption that has been made is a solar panel efficiency of 75 percent for a well designed unit. Each manufacturer makes his own claims and you must be the judge. The approach will be to calculate the square footage of panels required to offset the losses. Once the heat losses are covered, you can rely on the net heat gain from the sun shining directly on the pool surface. The collectable energy for the worst month in the season is developed from Table 5-1. A correction factor called a *radiation multiplier* is then applied for the same month to reflect the closest angle at which the collectors are mounted, either the latitude of your particular city or latitude minus 10°. You will find the latter is more efficient. Radiation multipliers may be calculated from Table 5-2. Lastly, the amount of energy required to offset losses is read from Fig. 5-9 for the appropriate pool area.

The mean daily solar radiation for the fringe months were obtained from Table 5-1. Since September is the lowest value, the figure of

Example:

Location		Indianapolis, Indiana
Latitude		39°
Desired swimming season		May through Sept.
Collector tilt (slope)		40°
Orientation		true south ± 20°
Minimum daytime air temperature		60° F
Mean daily solar radiation (langleys)		
	May	488
	Sept.	405
Pool surface area		800 sq. ft.

405 langleys must be used (1 langley = 3.687 Btu). This is the amount of radiation falling on a 1-square foot horizontal surface each day. As the panels are mounted at the latitude angle, a radiation multiplier of 1.2 applies (Table 5-2). For an 800 sq. ft. pool, losses of 625,000 Btu/day are indicated by Fig. 5-9.

Once the panel size is known, you should provide for the circulation of pool water through the panels at a rate of flow that produces the

Energy collected by panel surface
= langleys × 3.687 Btu/langley × radiation multiplier × panel efficiency
= 405 × 3.687 × 1.2 × 0.75
= 1344 Btu/sq ft - day

Energy required to offset pool losses
= 625,000 Btu/day

Panel surface required to offset pool loss

$$= \frac{625{,}000}{1344}$$

= 465 sq ft

highest efficiency. This rate is recommended by the panel manufacturer and is generally in the range of 2 to 4 gallons per minute/panel. The manufacturer should also tell you what the *pressure drop* will be across the panel at the flow rate specified. Sometimes this is called *head* and is shown in a curve like Fig. 5-10. One foot of water head equals 0.433 pounds per square inch (psi) of pressure. Or conversely,

Table 5-2. Effect of Tilt Angle on South-Facing Solar Panels for Various Latitudes.

Month	Degrees Latitude	Radiation Multiplier	
		Tilt = Latitude	Tilt = Latitude— 10°
Jan.	24 32 40 48	1.3 1.6 1.9 2.5	1.2 1.4 1.8 2.3
Feb.	24 32 40 48	1.2 1.3 1.5 1.8	1.1 1.3 1.5 1.7
Mar.	24 32 40 48	1.1 1.1 1.3 1.4	1.1 1.1 1.2 1.4
Apr.	24 32 40 48	1.0 1.0 1.0 1.1	1.0 1.0 1.1 1.1
May	24 32 40 48	0.9 0.9 0.9 0.9	1.0 1.0 1.0 1.0
June	24 32 40 48	0.8 0.8 0.8 0.8	0.9 0.9 0.9 0.9
July	24 32 40 48	0.9 0.9 0.9 0.9	1.0 1.0 1.0 1.0
Aug.	24 32 40 48	1.0 1.0 1.0 1.0	1.0 1.0 1.0 1.1
Sept.	24 32 40 48	1.1 1.1 1.2 1.4	1.1 1.1 1.2 1.4
Oct.	24 32 40 48	1.2 1.3 1.5 1.8	1.1 1.3 1.5 1.7
Nov.	24 32 40 48	1.3 1.5 1.9 2.4	1.2 1.4 1.7 2.2
Dec.	24 32 40 48	1.4 1.7 2.1 2.8	1.3 1.5 1.9 2.5

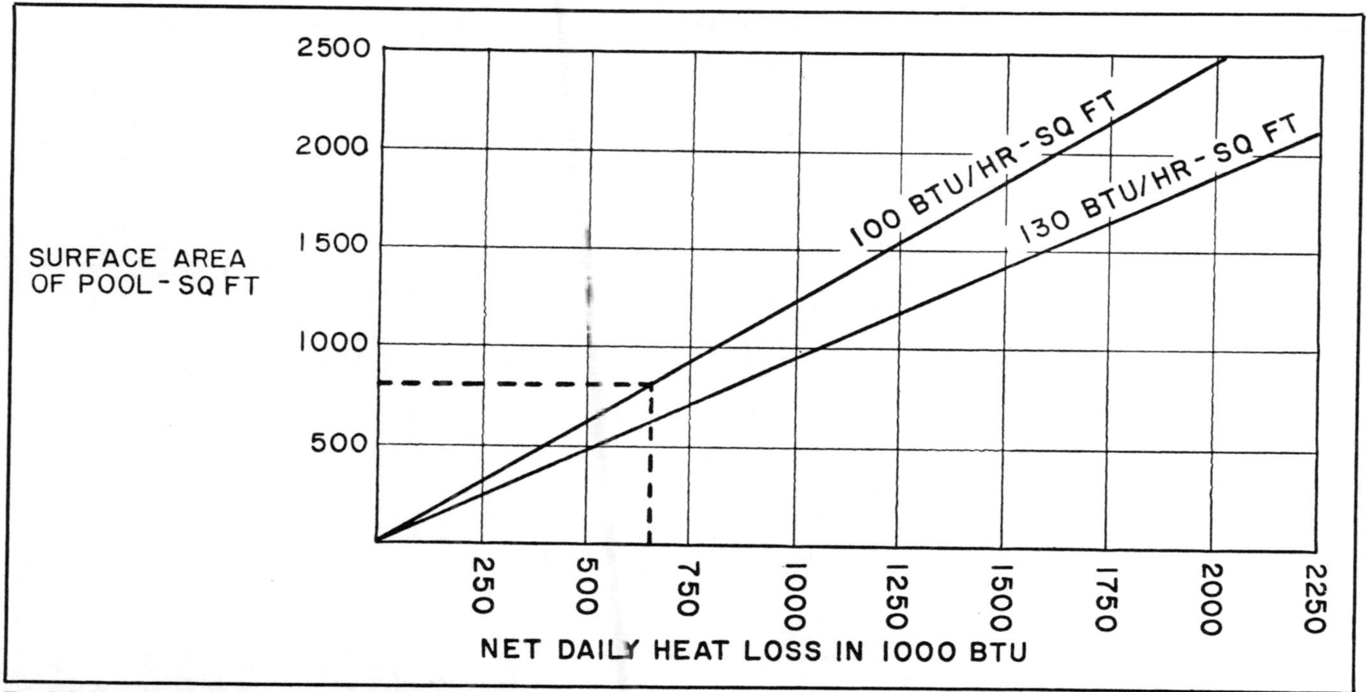

Fig. 5-9. Heat losses as a function of pool surface area.

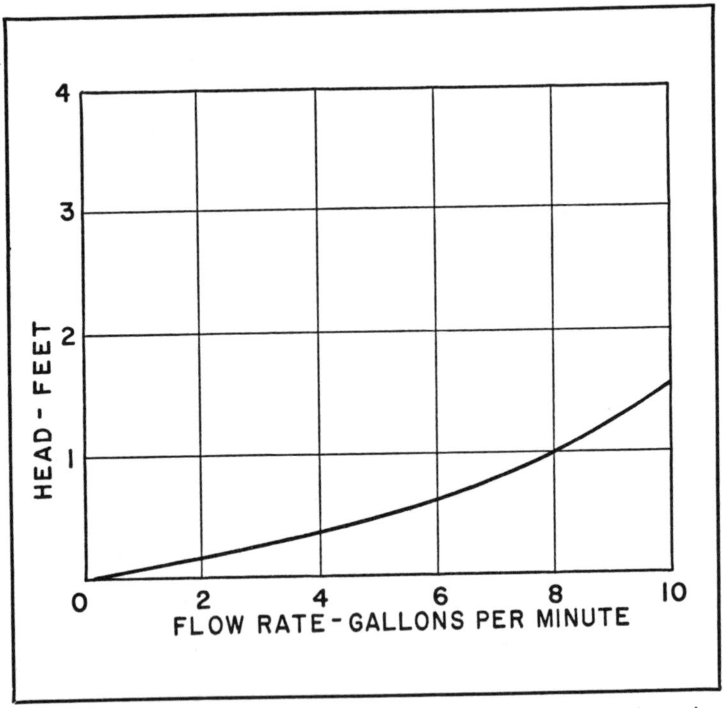

Fig. 5-10. Pressure drop at various flow rates for a typical solar pool panel.

$$\text{head (feet)} = 2.31 \times \text{pressure (psi)}$$

Regardless of the terminology used, we want the same result: namely, to determine if the filter pump now on your pool has enough is shown in Fig. 5-11. Again, it should be emphasized that these design suggestions are intended for relatively simple installations where the pipe runs are short and the roof space is adequate to capacity to handle the solar system in addition to the present load, or if a booster pump will be needed.

The pump controls the rate of water flow which is affected by the head, or pressure, of water it must pump against. Before adding solar heat, the circulation pump was sized by the pool contractor to maintain pool water filtration at a rate specified to meet local health and safety standards. Typically, this means moving the entire volume of water in the pool through the filter once every eight to twelve hours. For example, a pool holding 27,000 gallons circulating in 12 hours would require a flow rate of 38 gallons per minute. This is within the range of 20 to 40 gpm where most home pool systems operate.

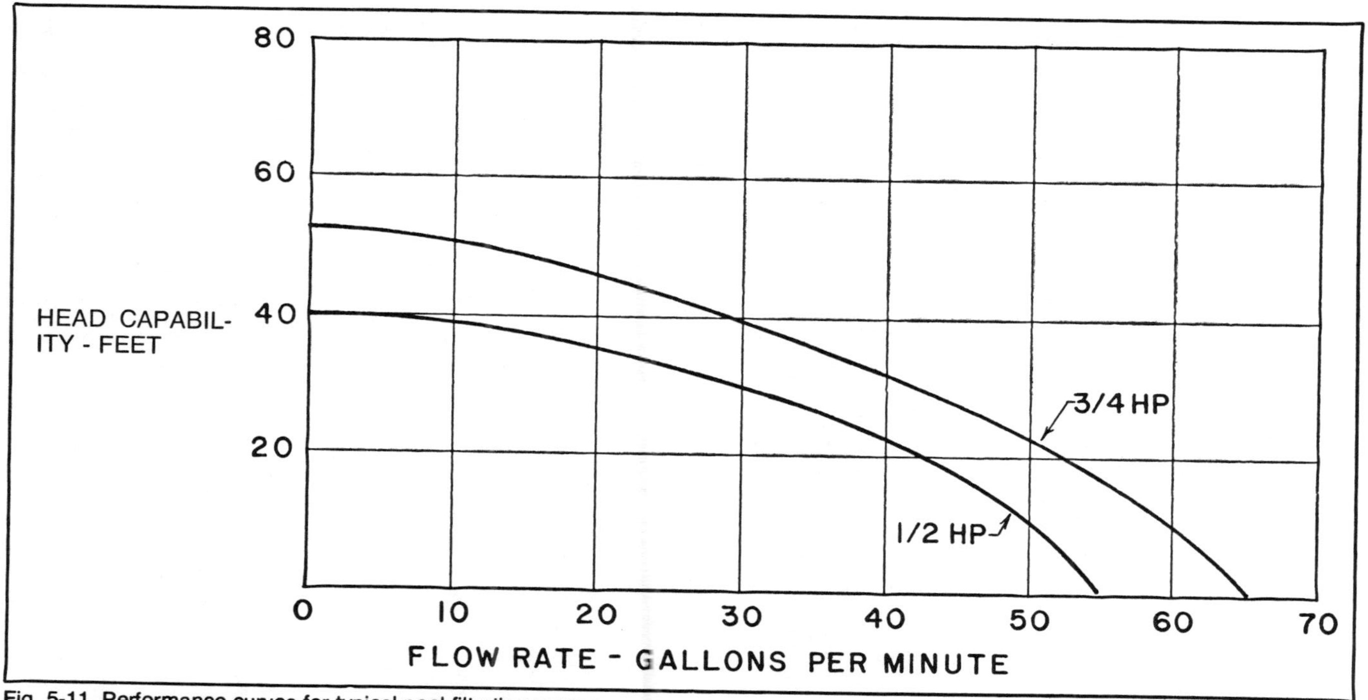

Fig. 5-11. Performance curves for typical pool filtration pumps.

When the solar system is added, additional requirements are imposed upon the pump to overcome frictional forces from the solar panels, valves, fittings, connecting piping, *plus* the vertical height to which the water must be raised by the pump. The total head is determined when all these factors are added together.

Performance curves for pumps show their head capability in feet of water for various flow rates. A typical pool filter pump curve install the panels in a single or, at most, two banks (Fig. 5-12). If panels must be divided between different roof surfaces or placed at varying heights and angles, the head calculations are more difficult and professional consultation is suggested. There is an excellent reference available from Solar Industries in Farmingdale, New Jersey, entitled "Solar Swimming Pool Heating System Hydraulics." It provides more detailed calculations and solutions to more complex installations.

The first step is to select a specific brand and model of panel. Once you determine its parameters (size, efficiency, flow rate, pressure drop, etc.), make a layout showing the way panels will be connected on the roof. The panel manufacturer should advise the maximum number of panels that can be connected in a single bank. (The total number of panels required was calculated previously.) In preparing the layout, there are some general rules you should follow:

- Connect each bank in parallel; bottom-to-bottom, top-to-top.
- Feed line should connect at the lowest point and the return line at the highest point in the system.
- Feed and return lines at opposite ends of a bank (called reverse return) provides a more balanced flow and is preferred.
- A vacuum relief valve should be placed at the highest point in the system (Fig. 5-11).
- Piping should slope so that all panels and lines will drain back to the feed line in the event of a power failure or when the solar panels have been bypassed by the diverting valve. *If you live in a freezing area this is absolutely essential* to prevent damage to the panels and lines.

Next, determine the vertical distance from the pool surface to the highest point on the return line once the panels are mounted. The pump must be capable of lifting water over this vertical distance, called the *static head.* Pump performance is determined from a curve supplied by the manufacturer plotting the head against flow rate. If you are unable to obtain one for your unit, an approximation may be made by noting the filter pump horsepower rating from the

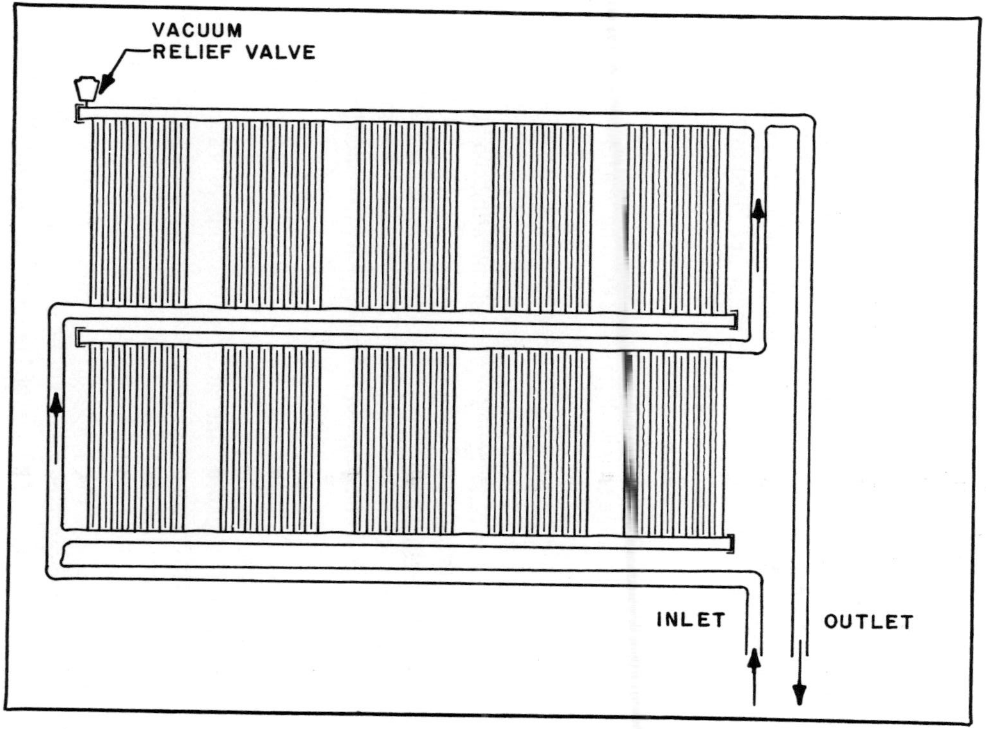

Fig. 5-12. Plumbing layout for two banks of solar panels on same roof surface.

nameplate. Use the family of curves from Fig. 5-13, and select the next lower rating if yours falls in between (e.g., if ¾ HP, use ½). If your static head is greater than the pump will deliver at zero flow rate, then you must use a booster pump. Don't cut this down to the last foot! As a safety factor, add 10 percent to the static head measured. Even if you pass this test, a booster may still be required as a specific flow rate must be met. This will be determined next.

The head requirement increases as water is forced to flow through the system due to frictional forces. A pressure drop is caused by fittings, valves, pipe, and the solar panels themselves. This loss is called the *dynamic head,* and must also be considered in calculating the pump size needed. Figures 5-14 and 5-15 show the effect of plastic piping and fittings. The dotted lines represent data points used in the next example. Once the pressure drop at different flow rates for the system you are considering is calculated, these points may be plotted along with the pump curve to determine the *operating point* of the system. The operating point is defined by the intersection of the pump and system curves. It is significant because it defines the rate of flow that can be expected through the solar panels. This rate must be within the range specified by the panel manufacturer to get the best performance from the solar system.

First you must establish a flow and pressure reading for your present circulation system without solar heating. Most systems have a pressure gauge on the filter; some even have a flow meter in the line. If these are in place, you need only to read them. If not, they should be installed. Referring to Fig. 5-4, the pressure gauge should be at point C and the flow meter at point A (or anywhere in the main pump line). Make a reading under normal operating conditions and with a clean filter. Adding the solar panels and associated piping will *increase* the pressure and *reduce* the system flow rate. The following example will illustrate how these calculations are made to determine if the existing pump has sufficient capacity to handle the solar addition.

Example: A homeowner has decided to add a solar system to his existing pool. A total panel area of 480 sq. ft. was estimated, and after reviewing literature supplied by various manufacturers, a 4′ × 10′ size panel was selected. A panel layout was sketched out showing two banks (Fig. 5-12). The highest point in the system is 30 feet above the pool surface (static head), and a motorized diverter valve will be used. The solar addition will require 120 feet of 1½-inch schedule 40 PVC pipe with 12 (large radius) fittings. Determine if a booster pump will be necessary.

Data Required	Value	Data Source(s)
Size of existing pump	1 HP	nameplate
Pump performance curve		Manufacturers' data (Fig. 5-13 used in example)
System flow without solar	30 gpm	measured
System pressure without solar	20 psi	measured
Total solar panel area	480 sq. ft.	calculated
Panel size selected	4′ × 10′	
Number of panels required	12	
Head loss per panel at measured flow using parallel connection	0.2 ft.	manufacturers' data (Fig. 5-10 used in example)
Specified flow rate per panel	2 gpm (min) 4 gpm (max)	manufacturers' data
Head loss from piping	6 ft.	Fig. 5-14
Head loss from fittings	8.4 ft.	Fig. 5-15

The present system flow of 30 gpm causes the following dynamic head losses due to the *solar* piping:

12 panels (0.2 ft. × 12) = 2.4 ft.
120 ft. of pipe (5 ft. × 1.2) = 6.0 ft.
12 fittings (0.7 ft. × 12) = 8.4 ft.
sub total 16.8 ft.

Additionally, the system pressure without a solar system measured to be 20 psi which, when converted to feet of head, becomes:

20 psi × 2.31 ft./psi = 46.2 ft.
total dynamic head at
30 gpm flow . . . . 63 ft

We now use a formula to calculate the system curve from the relationship between flow rate and head. The system head varies as the *square* of the change in flow rate:

$$\left(\frac{F_2}{F_1}\right)^2 = \frac{h_2}{h_1}$$

where:

F_1 = measured flow = 30 gpm
F_2 = final flow
h_1 = known head = 63 ft.
h_2 = final head

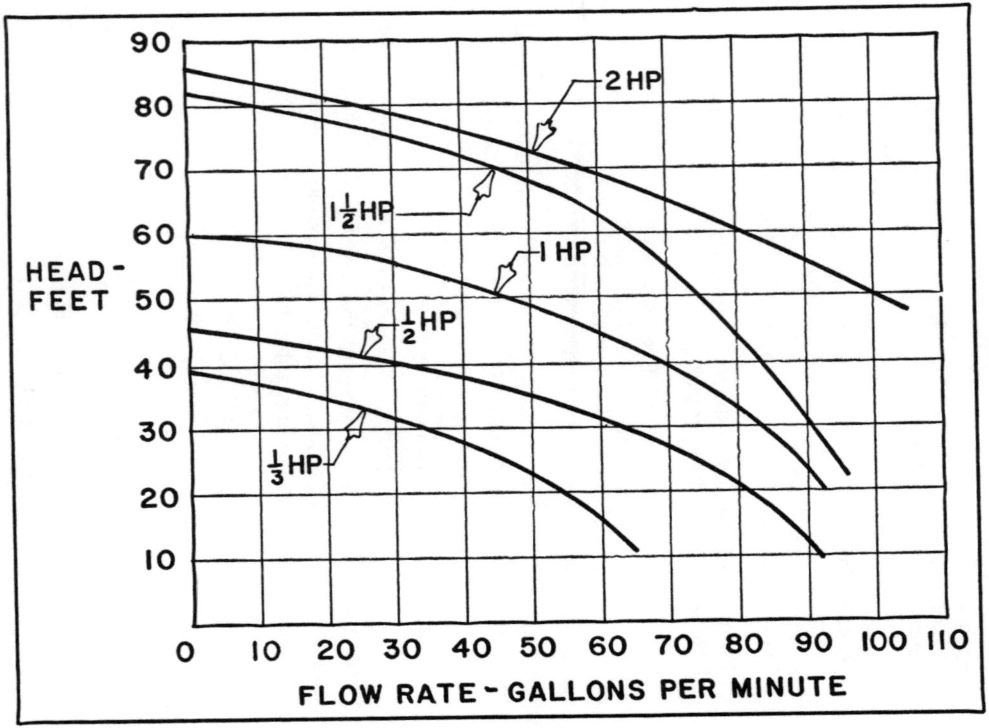

Fig. 5-13. Graph used to approximate static head capability based upon horsepower rating.

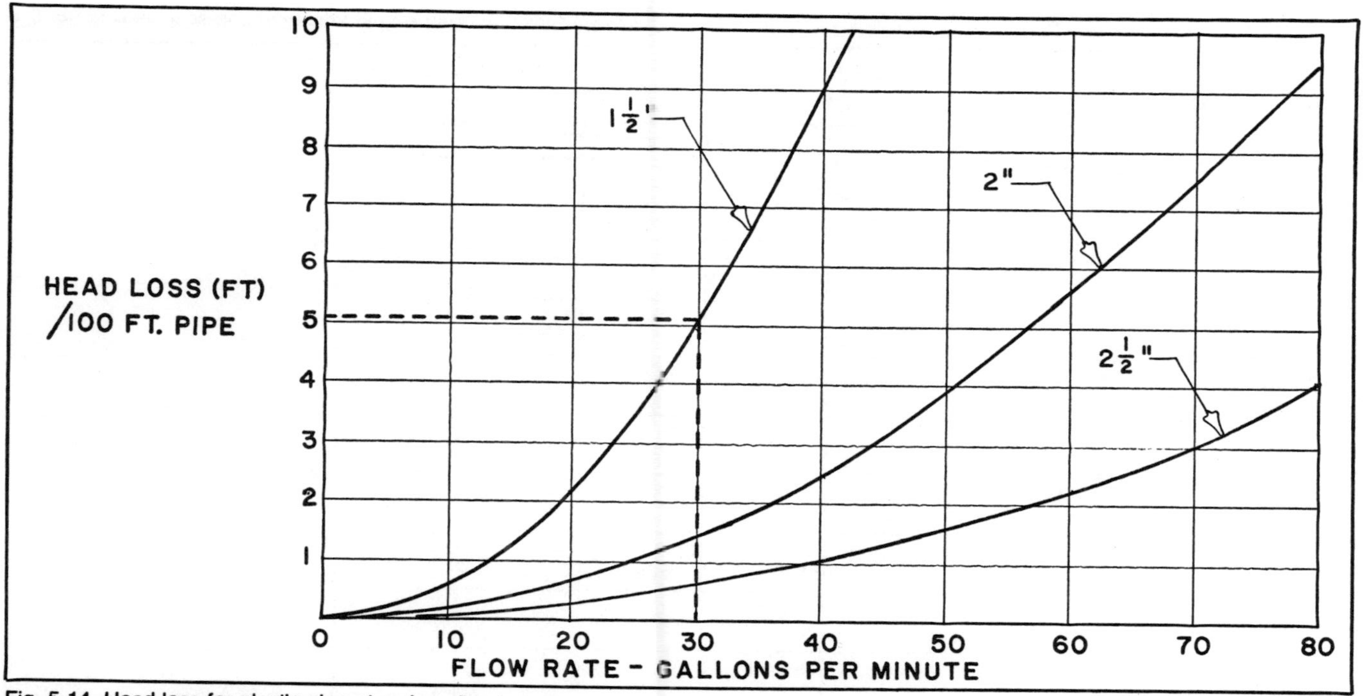

Fig. 5-14. Head loss for plastic pipe at various flow rates.

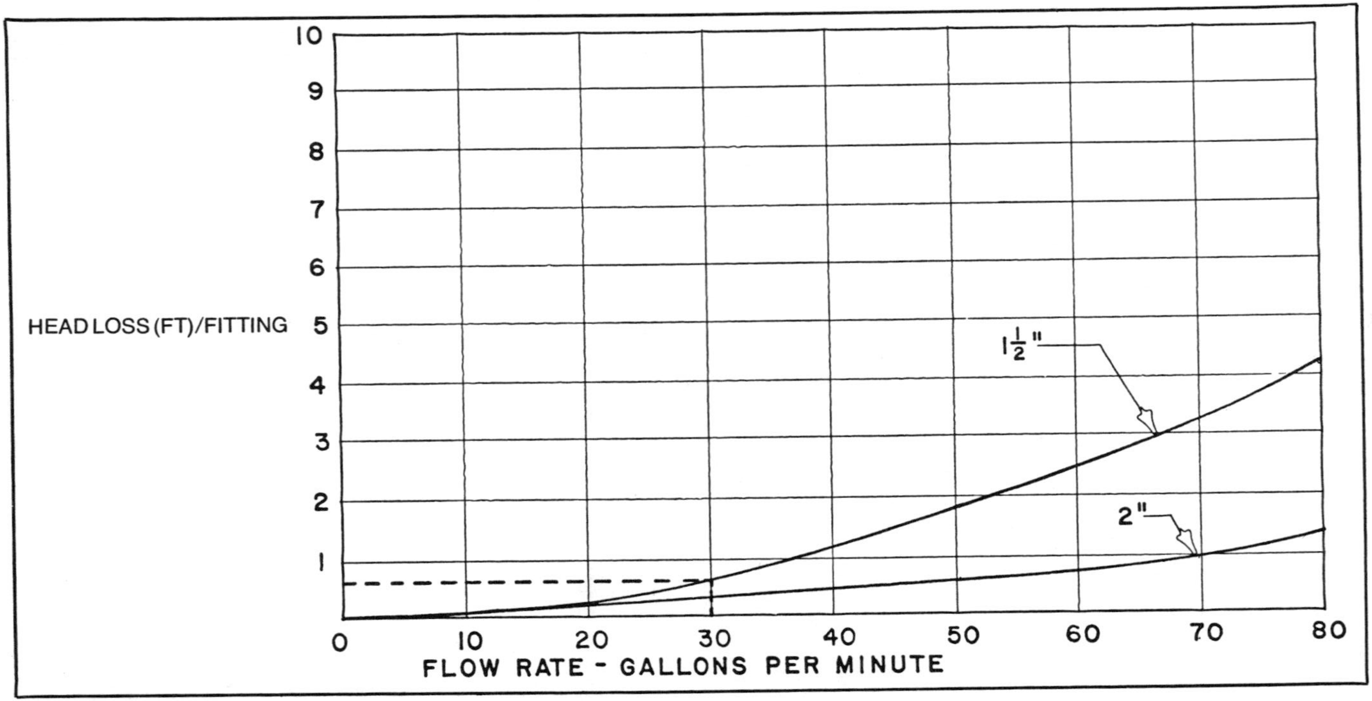

Fig. 5-15. Head loss due to friction of flow through plastic fittings.

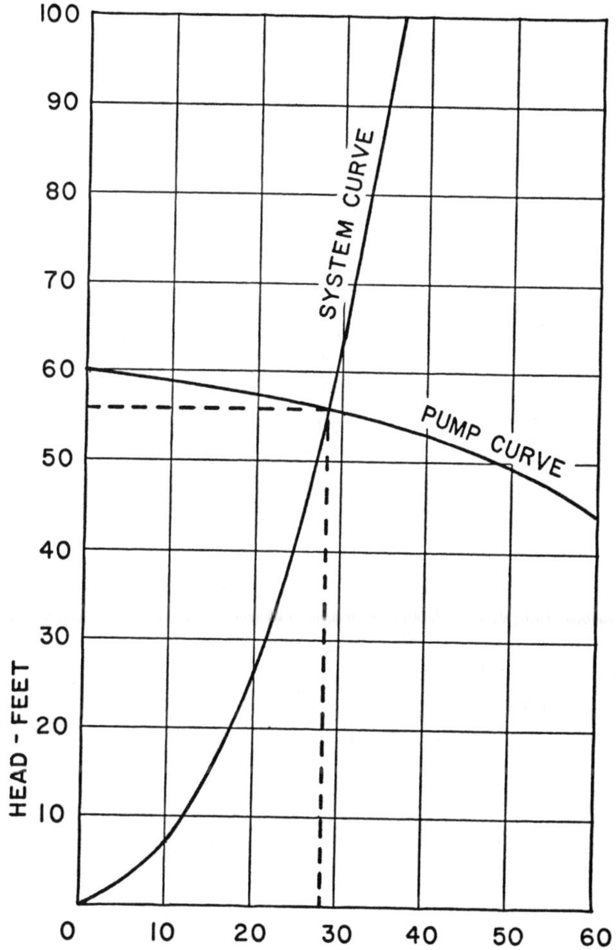

Fig. 5-16. Determination of the operating point used in the example.

Assume values for F_2 between zero and 40 and solve for the final head that corresponds to these values:

Starting with $F_2 = 10$

$$h_2 = \left(\frac{10}{30}\right)^2 \times 63$$
$$= 7$$

The coordinates for five points on the system curve are:

F_2	h_2
0	0
10	7
20	28
30	63
40	112

Figure 5-16 shows a plot of the system curve just developed against the 1 HP pump curve. The operating point (intersection) is at 28 gpm, suggesting that the addition of a solar system will reduce the normal measured flow by 2 gpm. At this rate, flow through each of the 12 solar panels connected in two parallel banks (Fig. 5-12) is 28/12 = 2.3 gpm, just above the minimum of 2 specified by the manufacturer. There is no problem in overcoming the static head of 30 feet because the pump has a lift capacity of close to 60 feet while the panels are being filled with water.

Although the 1 HP pump just makes the minimum flow specified, a dirty pool filter could cause the flow to fall below the recommended rate, and reduce panel efficiency. Consideration should be given, therefore, to adding a booster pump in this example.

Table 5-3. Factors for Economic Analysis.

1. Total Heating Requirement
2. Solar Fraction
3. Natural Gas Price (Current)
4. Inflation Rate
5. Useful Life
6. Market Discount Rate
7. State Tax Credit
8. Escalation Rate of Operating Costs
9. Escalation Rate of Maintenance Costs
10. Escalation Rate of Fuel Costs
11. Solar Maintenance Costs ($/Yr)
12. Solar Operating Costs ($/Sq Ft/Yr)

A LOOK AT ECONOMICS

You should now have a good understanding of how heat losses from a swimming pool occur, how they can be minimized, and the technical side of solar pool heating systems. Now you may evaluate the economics of installing this equipment. When consulting on large installations, I use a proprietary computer program with the 12 input parameters shown in Table 5-3. You may not wish to get into this much detail for home pools, but many of the same decisions apply. For example:

- What will natural gas cost in the future?
- How much gas will be saved?
- How long will the solar equipment last and how much will it cost to install and maintain?

In any case, remember that the energy you save with solar devices is equivalent to tax free money; the interest earned from treasury bills is not.

Wind Energy Incentive Programs

July 1980

Alaska
Bob Shipley
State of Alaska Division of Energy and Power Development
338 Denali Street
Anchorage, AK 99501
(907) 276-0508

Arkansas
Nick Brown
Arkansas Department of Energy
3000 Kavanaugh Boulevard
Little Rock, AR 72205
(501) 371-1370

California
John L. Obermeir
Office of Appropriate Technology
1322 0 Street
Sacramento, CA 95814
(916) 322-8519

Colorado
Peggy Wren
Colorado Office of Energy Conservation
1600 Downing Street
Denver, CO 80218
(303) 839-2507

Connecticut
Leah Haygood
Office of Policy and Management
80 Washington Street
Hartford, CT 06115
(203) 566-3394

Hawaii
Eugene M. Grabbe
Department of Planning and Economic Development
P.O. Box 2359
Honolulu, HI 96804
(808) 548-4195

Idaho
Megan Morgan
Idaho Office of Energy
State House
Boise, ID 83720
(208) 334-3800

Illinois
Gary Mielke
Institute of Natural Resources
Solar Section
325 West Adams Street
Springfield, IL 62706
(217) 785-2433

Indiana
Mary Failey
Program Coordinator
Indiana Solar Office
440 North Meridian Street
Indianapolis, IN 46204
(317) 232-8940

Iowa
Philip Svanoe
Iowa Energy Policy Council
215 East 7th Street
Des Moines, IA 50319
(515) 281-4420

Kansas
David Martin
Coordinator, Solar and
Wind Program
Kansas Energy Office
214 West 6th Street
Topeka, KS 66603
(913) 296-2496

Maine
Vincent DiCara
Office of Energy Resources
55 Capitol Street
Augusta, ME 04330
(207) 289-3811

Massachusetts
Phyllis Gardiner
Executive Office of
Energy Resources
73 Tremont Street
Boston, MA 02108
(617) 727-4732

Michigan
Rich Whale
Michigan Energy Administration
Box 30228
Lansing, MI 48909
(517) 373-6430

Minnesota
John Cutty
Minnesota Energy Office
160 Kellogg Boulevard
St. Paul, MN 55101
(612) 297-2327

Montana
Jim Iverson
Department of Natural
Resources
32 South Ewing
Helena, MT 59601
(406) 449-4624

Nevada
Bob Loux
Department of Energy
1050 E. Williams Street
Suite 405
Carson City, NV 89710
(702) 885-5157

New Hampshire
Julie Eades
State Energy Office
2½ Beacon Street
Concord, NH 03301
(603) 271-2711

New Jersey
Joe Oliva
Department of Energy
101 Commerce Street
Newark, NJ 07102
(201) 648-3572

New York
Paul Wendelgass
New York State Energy Office
Agency Building No. 2
Empire State Plaza
Albany, NY 12223
(518) 473-8251

North Dakota
John Conrad
Office of Energy Management and Programs
1533 North 12th Street
Bismarck, ND 58501
(701) 224-2250

Ohio
Leon Winget
Ohio Department of Energy
30 E. Broad Street - 34th Floor
Columbus, OH 43215
(614) 466-6797

Oregon
David Philbrick or Don Bain
Oregon Department of Energy
Labor and Industry Building - Room 111
Salem, OR 97310
(503) 378-5584

South Dakota
Steve Wagman
Office of Energy Policy
Capitol Lake Plaza
Pierre, SD 57501
(605) 773-3603

Tennessee
Eugene Edmunds
Tennessee Energy Authority
Suite 707
Capitol Boulevard
Nashville, TN 37219
(615) 741-2994

Texas
Bob Avant
Texas Energy Advisory Council
411 West 13th Street
Austin, TX 78701
(512) 475-5588

Vermont
Richard Cambio
Vermont Energy Office
State Office Building
Montpelier, VT 05602
(802) 828-2393

Virginia
Temple Bayliff
Division of Energy
Commonwealth of Virginia
310 Turner Road
Richmond, VA 23225
(804) 745-3245

Wisconsin
Bonnie Albright
Office of State Planning and Energy
1 Wilson Street
Madison, WI 53702
(608) 266-9861

B

Electric Wind Machine Specifications

Name—Model Number	Size (rated kW)	Rotor Diameter (feet/meters)	No. Blades	Rated Speed mph/ m/s	Cut-in Speed mph/ m/s
Mehrkam Energy Development Co. (USA)—440	40.00	40.0/12.2	6	25/11.2	7/2.2
Jay Carter (USA)—Mod 25	25.0	32.0/ 9.8	2	25/11.2	N/A
Wind Engineering (USA)—Windgen 25	25.0	38.0/11.6	3	25/11.2	N/A
Tumac Industries (USA)—10-meter*	20.0	32.8/10.0	3	N/A	N/A
High-Speed Generator					
Grumman Energy Systems (USA—Windstream 25-B	20.000	25.0/ 7.62	3	29/12.9	8/3.6
Wind Power Systems, Inc. (USA)—Storm Master	18.0	32.8/10.0	3	24/10.7	8/3.6
Environmental Energies (USA)—HWT 15	15.0	24.0/	8	25/11.0	8.7/3.9
Environmental Energies (USA)—HWT 12	12.0	22.0/	8	25/11.0	8.7/3.9
Astral-Wilcon (USA)—AW 10-B	10.0	26.0/ 7.7	3	N/A	N/A
Millville Windmills (USA)—10-3-IND	10.00	25.0/ 7.60	3	25/11.0	9/4.0
Elektro (Switz)—WVG 120G	10.0	21.7/ 6.6	3	31/13.8	7/3.2
Environmental Energies (USA)—HWT 9	9.0	20.0/	8	25/11.0	8.7/3.9
Environmental Energies (USA)—HWT 6	6.0	18.0/	8	25/11.0	8.7/3.9
Elektro (Switz)—WVG 50G	5.4	16.4/ 5.0	3	27/12.2	7/3.2
Tumac Industries (USA)—10-meter*	5.0	32.8/10.0	3	N/A	N/A
Low-Speed Generator					
Tumac Industries (USA)—5-meter*	5.0	16.4/ 5.0	3	N/A	N/A
High-Speed Generator					
Independent Energy Systems, Inc. (USA)—Skyhawk	4.000	15.0/ 4.6	3	23/10.1	7-8/3.1-3.6

Aerowatt S.A. (France)—4100FP7G	4.100	32.7/10.0	2	16/7.2	7/3.1
Dakota Wind & Sun, Ltd. (USA)—4 kW	4.000	14.0/ 4.3	3	27/12.0	8/3.6
Power Group Int'l Corp. (USA)—Hummingbird 4000	4.0	14.0/ 4.3	3	22/10.0	6-8/2.7-3.6
Elektro (Switz)—WV 35G	3.8	14.5/ 4.4	3	27/12.2	7/3.2
Dynergy (USA)—5-meter*	3.300	15.0/ 4.6	2 (Darrieus)	24/10.7	10/4.4
Hinton Research (USA)	3.0	N/A	N/A	N/A	N/A
Product Development Institute (USA)—Wind Heat System (prototype)	3.000	13.6/ 4.15	3	27/12.0	12/5.4
Product Development Institute (USA)—Wind Jennie (prototype)	3.000	12.5/ 3.8	3	25/11.1	12/5.4
Kedco, Inc. (USA)—1620	3.000	16.0/ 4.88	3	25/11.1	11/4.9
Altos, The Alternate Current (USA)—BWP-12A	2.200	8.0/ 2.4	24	38/17.1	8/3.6
Elektro (Switz)—WV 25G	2.2	12.0/ 3.6	2	27/12.2	7/3.2
Pinson Energy Corporation (USA)—C-2E*	2.000	12.0/ 3.66	3	24/10.7	7/3.1
Altos, The Alternate Current (USA)—BWP-12B	2.000	11.5/ 3.5	N/A	28/12.5	8/3.6
American Wind Turbine, Inc. (USA)—16-foot	2.000	15.3/ 4.7	48	20/9.0	10/4.5
Bertoia Studio (USA)—A.P.S.	2.0	18.0/ 5.49	3	18/8.0	5/2.2
Dunlite Electrical Products (Australia)—81/002550 (Standard Model)	2.000	13.5/ 4.10	3	25/11.0	8/3.6

*Vertical axis.

Name—Model Number	Size (rated kW)	Rotor Diameter (feet/meters)	No. Blades	Rated Speed mph/ m/s	Cut-in Speed mph/ m/s
Kedco, Inc. (USA)—1210	2.000	12.0/ 3.65	3	26/11.6	11/4.9
Kedco, Inc. (USA)—1610	2.000	16.0/ 4.88	3	22/10.0	10/4.4
North Wind Power Company (USA)—Eagle 2 kW-32 V	2.000	13.6/ 4.15	3	22/10.0	8/3.6
North Wind Power Company (USA)—Eagle 2 kW-110V	2.000	13.6/ 4.2	3	22/10.0	8/3.6
Sencenbaugh (USA)—2000/115	2.0	12.0/ 3.65	3	26/11.6	N/A
Whirlwind Power Company (USA)—Model A	2.000	10.0/ 3.0	2	25/11.2	8/3.6
Winflo (Canada)—Wingen Mod 2000	2.0	10.0/ 3.0	2	25/11.2	N/A
Kedco, Inc. (USA)—1605	1.900	16.0/ 4.88	3	20/8.9	7/3.1
Enertech Corp. (USA)—1500	1.500	13.2/ 4.0	3	22/9.8	8/3.6
Altos, The Alternate Current (USA)—BWP-8A	1.500	8.0/ 2.4	24	28/12.5	8/3.6
Tumac Industries (USA)—5-meter* Low-Speed Generator	1.50	16.4/ 5.0	3	N/A	N/A
Aero Power Systems, Inc. (USA)—SL1500	1.430	10.0/ 3.05	3	25/11.2	6/2.7
Elektro (Switz)—WV 15G	1.200	10.0/ 3.0	2	27/12.2	7/3.2
Kedco, Inc. (USA)—1200	1.200	12.0/ 3.65	3	22/10.0	7/3.1
Kedco, Inc. (USA)—1205	1.200	12.0/ 3.65	3	22/10.0	8/3.6
Kedco, Inc. (USA)—1600	1.200	16.0/ 4.88	3	17/7.6	7/3.1
Wind Titan	1.200	14.0/ 4.3	2	25/11.2	7.5/3.4
Aerowatt S.A. (France)—1100PF7G	1.125	16.7/ 5.09	2	16/7.2	7/3.1

American Wind Turbine, Inc. (USA)—12-foot	1.000	11.5/ 3.5	36	20/9.0	10/4.5
Dunlite Electrical Products (Australia)—82/002550 (High Wind Speed Model)	1.000	10.0/ 3.1	3	37/16.5	14/6.3
Sencenbaugh Wind Electric (USA)—1000	1.000	12.0/ 3.65	3	23/10.3	6/2.7
Aerolectric Company (USA)—C-9/C-90 Wind Wizard	0.[illegible]00	9.0/ 2.75	3	26/11.6	9/4.0
American Wind Turbine, Inc. (USA)—8-foot	0.500	7.6/ 2.3	24	20/9.0	10/4.5
Elektro (Switz)—WV 05G	0.500	8.2/ 2.5	2	19/8.5	7/3.2
Sencenbaugh Wind Electric (USA)—500	0.500	6.0/ 1.8	3	24/10.8	10/4.5
Winflo (Canada)—Wingen Mod 500	0.500	6.0/ 2.0	3	25/11.2	7/3.0
Megatech Corp. (USA)—Model WIP-A1	0.400	6.0/ 2.0	2	27/12.2	6/2.7
TWR Engerprises (USA)	0.500	10.0/ 3.05	2	25/11.2	8.5/3.8
Sencenbaugh Wind Electric (USA)—400-14HDS	0.400	7.0/ 2.13	3	20/9.0	9/4.0
Aerowatt S.A. (France)—300FP7G	0.350	10.7/ 3.26	2	16/7.2	7/3.1
Dragonfly Wind Electric (USA)	0.240	8.0/ 2.4	4	20/9.0	10/4.5
Winco-Division of Dynatechnology (USA)—1222H	0.200	6.0/ 1.83	2	23/10.3	7/3.1
Aerowatt S.A. (France)—150FP7G	0.130	6.7/ 2.04	2	16/7.2	7/3.1
Winflo (Canada)—Wingen Mod 30	0.030	3.0/ 1.0	10	25/11.2	N/A
Aerowatt S.A. (France)—24FP7G	0.028	3.3/ 1.00	2	16/7.2	7/3.1
Zephyr Wind Dynamo Tetrahelix S	0.007	2.0/ 0.61	2	25/11.2	12/5.4

*Vertical axis.

Sources

COMMERCIAL SOLAR CELL MODULE MANUFACTURERS (Flat, Nonconcentrator type)

Applied Solar Energy Corporation
15251 East Don Julian Road
City of Industry, CA 91746
Attn: Wayne C. Stevenson, Sales Manager
(213) 968-6581

Arco Solar, Inc.
20554 Plummer Street
Chatsworth, CA 91311
Attn: Jack Cotter, Sales Manager
(213) 998-2482

Motorola, Inc.
Solar Energy Department
Phoenix, AZ 95008
Attn: Bob Hammond, Marketing Manager
(602) 244-5459

Photowatt International, Inc.
21012 Lassen Street
Chatsworth, CA 91311
Attn: Sanjeev Chitre, Vice President, Director, Marketing
(213) 882-4100

SES Incorporated
Tralee Industrial Park
Newark, DE 19711
Attn: Greg Love, Manager of Industrial Sales
(302) 731-0990

Silicon Sensors, Inc.
Highway 18 East
Dodgeville, WI 53533
Attn: Robert Bachner, President, Marketing Manager
(608) 935-2707

Solarex Corporation
1335 Piccard Drive
Rockville, MD 20850
Attn: Ted Blumenstock, Director of Marketing
(301) 948-0202

Solar Power Corporation
20 Cabot Road
Woburn, MA 01801
Attn: Bill Brusseau, Marketing Manager
(617) 935-4600

Solec International, Inc.
12533 Chadron Avenue
Hawthorne, CA 90250
Attn: Robert W. Crawford, Vice President, Marketing
(213) 970-0065

Solenergy Corporation
23 North Avenue
Wakefield, MA 01880
Attn: Robert W. Willis, President, Marketing Manager
(617) 246-1855

Sollos, Inc.
2231 S. Carmelina
Los Angeles, CA 90064
Attn: Dr. Milo Macha, President, Marketing Manager
(213) 820-5181

Tideland Signal Corporation
4310 Directors Road
P.O. Box 52430
Houston, TX 77052
Attn: Carl Kotila, Environmental Energy Manager
(713) 681-6101

ELECTRIC WIND MACHINE MANUFACTURERS

Aeroelectric
13517 Winter Lane
Cresaptown, MD 21502
(609) 547-3488
Kevin Moran
Wind Wizard

Aero Power
2398 Fourth Street
Berkeley, CA 94710
(415) 848-2710
John Harold or Tom Cummins

Aerowatt, S.A.
c/o Automatic Power, Inc.
P.O. Box 18738
Houston, TX 77023
(713) 228-5208
Robert Dodge or Ernest Tindle

Alcoa
Alcoa Laboratory
Alcoa Center, PA 15069
Darrieus Systems
(412) 339-6651

Altos: The Alternate Current
P.O. Box 905
Boulder, CO 80302
(303) 442-0885
Edward Gitlin

American Wind Turbine, Inc.
1016 East Airport Road
Stillwater, OK 74074
(405) 377-5333
Nancy Thedford, Office Manager

Astral/Wilcon
P.O. Box 291
Milbury, MA 01527
Bill Stein or Marcus Sherman

Bergey Wind Power Co.
2001 Priestly Avenue
Norman, OK 73069
(405) 364-4212
Mike Bergey

Bertoia Studio
644 Main Street
Bally, PA 19503
(215) 845-7096
Val O. Bertoia
A.P.S

Jay Carter Enterprises
P.O. Box 684
Burkburnett, TX 76354

Chalk Wind Systems
P.O. Box 446
St. Cloud, FL 32769
(305) 892-7338

Dakota Wind & Sun, Ltd.
P.O. Box 1781
811 First Avenue, N.W.
Aberdeen, SD 57401
(605) 229-0815
Paul Biorn or Orv Lynner

Dragonfly Wind Electric
P.O. Box 155
Elk, CA 95432
(707) 877-3474

Dunlite Electrical Products Co.
Enertech Corporation
P.O. Box 420
Norwich, VT 05055
(802) 649-1145
Ned Coffin or Robert Sherwin

Dynergy Corporation
P.O. Box 428
1269 Union Avenue
Laconia, NH 03246
(603) 524-8313
Robert Allen

Enertech Corporation
P.O. Box 420
Norwich, VT, 05055
(802) 649-1145
Ned Coffin or Robert Sherwin

Environmental Energies, Inc.
Front Street
Copemish, MI 49625
(616) 378-2000
Timothy J. Horning

Grumman Energy Systems
4175 Veterans Memorial Highway
Ronkonkoma, NY 11779
(516) 575-6205
Ed Diamond or Ken Speiser

Hinton Research
417 Kensington
Salt Lake City, UT 84115
(801) 487-3896

Independent Energy Systems, Inc.
6043 Sterrenttania Road
Fairview, PA 16415
(814) 833-3567

John D'Angelo
Skyhawk

Jacobs Wind Electric Co., Inc.
Route 13, Box 722
Fort Myers, FL 33908
(813) 481-3113
(813) 334-0339
Paul Jacobs or M.L. Jacobs

Kaman Aerospace
Old Windsor Road
Bloomfield, CT 06002
(203) 242-4461

Kedco, Inc.
9016 Aviation Boulevard
Inglewood, CA 90301
(213) 776-6636
Wind Program Manager

McDonnell Aircraft Co.
P.O. Box 516
St. Louis, MI 63166
(314) 232-3575
Robert V. Brulle

Megatech Corporation
29 Cook Street
Billerica, MA 01866
(617) 273-1900

Mehrkam Energy Development Co.
179 East Road #2
Hamburg, PA 19526
(215) 562-8856
Terrance or Helena Mehrkam or Karen Votyas
EDC 440

Millville Windmills & Solar Equipment Company
P.O. Box 32
10335 Old Drive
Millville, CA 96062
(916) 547-4302
Devon Tassen

North Wind Power Company
P.O. Box 315
Warren, VT 05674
(802) 496-2955
Don Mayer

Pinson Energy Corporation
P.O. Box 7
Marstons Mills, MA 02648
(617) 477-2913
Herman Drees

Power Group International Corp.
Suite 106
13315 Stuebener Airline Road
Houston, TX 77014
(713) 444-5000
Michael Graham
Hummingbird

Product Development Institute
508 South Byrne Road
Toledo, OH 43609
(419) 382-0282
Tom Nichols
Wind Jennie

Sencenbaugh Wind Electric
P.O. Box 11174
Palo Alto, CA 94306
(415) 964-1593
Jim Sencenbaugh

Tumac Industries
650 Fort Street
Colorado Springs, CO 80915
(303) 596-4400

TWR Enterprises
Sun-Wind-Home Concepts
72 West Meadow Lane
Sandy, UT 84070
Tom W. Rentz
Wind Titan

Winco
Division of Dyna Technology
7850 Metro Parkway
Minneapolis, MN 55420
(612) 853-8400
Len Attema
Windcharger

Windworks, Inc.
Box 329, Route 3
Mukwonago, WI 53149
(414) 363-4408/9
Hans Meyer

Whirlwind Power Company
2458 W. 29th Avenue
Denver, CO 80211
(303) 477-6436
Elliott Bayly

Wind Power Systems, Inc.
P.O. Box 17323
San Diego, CA 92117
(714) 452-7040
Ed Salter
Storm Master

Winflo Power Ltd.
90 Esna Drive, Unit 15
Markham, Ontario
Canada L3R, 2R7
Wingen

W.T.G. Energy Systems, Inc.
P.O. Box 87
1 LaSalle Street
Angola, NY 14006
(716) 549-5544
Al Wellikoff

Zephyr Wind Dynamo
Company
P.O. Box 241
21 Stamwood Street
Brunswick, ME 04011
(207) 725-6534
Bill Gillette

MECHANICAL WIND MACHINE MANUFACTURERS

Aermotor Division
Valley Industries
P.O. Box 1364
Conway, AR 72032
(501) 329-9811
Stan Anderson

American Wind Turbine, Inc.
1016 East Airport Road
Stillwater, OK 74074
(405) 223-4026
Nancy Thedford, Office Manager

Bowjon
2829 Burton Avenue
Burbank, CA 91504
(213) 846-2620

Dempster Industries, Inc.
P.O. Box 848

Beatrice, NB 68310
(402) 223-4026
Roy Smith

Dynergy Corporation
P.O. Box 428
1268 Union Avenue
Laconia, NH 03246
(603) 524-8313
Bob Allen

Heller-Aller Company
Perry & Oakwood Streets
Napoleon, OH 43545
Manufacturing of Baker Windmills
(419) 592-1856
James Bradner or Charles Buehrer

Sparco (Denmark)
c/o Enertech, Inc.
P.O. Box 420
Norwich, VT 05055
(802) 649-1145
Edmund Coffin at Enertech

Wadler Manufacturing Co., Inc.
Route 2, Box 76
Galena, KS 66739
(316) 783-1355
Jerry Wade

WIND MACHINE DEALERS /DISTRIBUTORS

Aermotor
1243 Majesty Drive
Dallas, TX 75247
(214) 634-1950
Mr. William E. Barney

Aermotor
2385 South Cherry
Fresno, CA 93706
(209) 486-7200

Aermotor
900 Nabco Avenue
P.O. Box 1321
Conway, AR 72032
(501) 329-2969
Miles Patten

Aermotor
6448 Warren Drive
Norcross, GA 30093
(404) 449-1840

Aermotor
518-M North Douglas Avenue
Altamonte Springs, FL 32701
(305) 862-0171

Aermotor
2421 West Main
Ft. Wayne, IN 46808
(219) 432-2595

Aermotor
4655 Colt Road
Rockford, IL 61109
(815) 874-9502

Aermotor
801 Howard Street
Omaha, NB 68102
(402) 341-1716
Marvin Vesenik

Aermotor
8105 Lewis Road
Minneapolis, MN 55427
(612) 544-4106

Aermotor
2803 South Longview Drive
Middletown, PA 17057
(717) 939-9311

Aermotor
1575 Avon Street Extended
Charlottesville, VA 22901
(804) 977-0445

Alternate Energy Systems
150 Sandwich Street
Plymouth, MA 02360
Elektro, Dunlite, Winco
(617) 747-0771

Automatic Power, Inc.
P.O. Box 18738
Houston, TX 77023
(713) 228-5208
Robert Dodge
Aerowatt

Avant Garde Entrepreneurs
1202 15th Street
Wheatland, WY 82201
(307) 322-3566
Dana V. West
Dakota Wind & Sun, Millville

Dean Bennet Supply Company
4725 Lipan Street
Denver, CO 80211
(303) 433-8291
Deana Bennet
Aermotor, Dunlite, Winco

Bergey Wind Power Co.
2001 Priestly Avenue
Norman, OK 73069
(405) 364-4212
Mike Bergey

Booger County Enterprises
Star Route
Witter, AR 72766

Chalk Wind Systems
P.O. Box 446
St. Cloud, FL 32769
(305) 892-7338

Chinook Wind Power, Inc.
1084 W. 101st Place
Denver, CO 80221
(303) 451-0660
Alice Unger
Aero Power, Sencenbaugh, Dakota Wind & Sun

Clean Energy Products
3534 Bagley, N.
Seattle, WA 98103
(206) 633-5505
Ed Kennel
Reconditioned Older Models, Kedco, Sencenbaugh, Windcharger

Cojo Wind Company
Hollister Ranch #6
Gaviota, CA 93017
(805) 963-5248
David A. Kay

Coulson Wind Electric
RFD 1, Box 225
Polk City, IA 50226
(515) 547-3488
R. Coulson
Reconditioned Older Models, Winco, Winpower

Crowdis Conservers
RR 3, MacMillan Mt.

Cape Breton, Nova Scotia
Canada BOE 1BO
Daniel Atkins
North Wind

Edmund Scientific Company
380 EDS Corp. Building
101 East Gloucester Pike
Barrington, NJ 08007
(609) 547-3488
Robert F. McKelvey
Aerolectric (Wind Wizard)

Empire Energy Development Corp.
3371 West Hampden Avenue
Englewood, CO 80110
(303) 789-1363
David L. Flook
Altos, Winco

Energy Alternatives
52 French King Highway
Greenfield, MA 01301
(413) 773-5175
Frank Kaminsky or
Klaus Kroner
Elektro, Dunlite, Winco, Sencenbaugh

Energy—2000
Route 800, RFD #3
Winstead, CT 06098
(203) 379-5185
(413) 528-3440
Robert Hebert
Reconditioned Older Models

Enertech Corporation
P.O. Box 420
Norwich, VT 05055
(802) 649-1145
E. Coffin
Dunlite, Winco, Sencenbaugh, Sparco

Environmental
Resource Group
Box 3A, RD 2
Williston, VT 05495
(802) 879-0511
(802) 878-4000
Perry Kleine
North Wind, Sparco

Fenton's Feeders
Route 1, Box 124
Arcadia, FL 33821
(817) 494-2727
Catherine Fenton, Bill Autry or Nell Gammage
Aermotor (water pumping)

4—Winds of Alaska
5100 Vi Street
Anchorage, AK 99507

Independent Energy Company
314 Howard Avenue
Swarthmore, PA 19081
(617) 368-6992
Kendall B. Hampton
Elektro, Dunlite, Winco

Independent Energy Systems
6043 Sterrettania Road
Fairview, PA 16415
(814) 833-0829
John D'Angelo
Skyhawk, Dakota Wind & Sun, Reconditioned Older Models

Jacobs Wind Electric Company, Inc.
Route 13, Box 722

Fort Myers, FL 33908
(813) 481-3113
(813) 334-0339
M.L. Jacobs or Paul Jacobs

Kemah Power Company
Box 776
633 Pleasant Avenue
Saugatuck, MI 49453

Kramco
P.O. Box 1536
Allentown, PA 18105
(215) 437-6758
Reconditioned Older Models, Winco

Laholms Motor & Bilelektriska A/B
Export Office & Information
Wind Kraft A.E.S.C.
Box #104
S-312 01 Laholm, Sweden
Int. Phone: 00946-43020371
Cable: WERKOMP
Eric Alkstrad
Wind Kraft A.E.S.C.

Living Energy Consultants
131 Beaver Avenue
Colorado Springs, CO 80906

Makia Ocean Engineering
Box 1194
Kailua, Oahu, Hawaii 96734
(808) 259-5904
(808) 259-5722
Henry Horn
Dunlite

Massawippi Wind Electric Co.
RR 3
Ayer's Clif, Quebec
Canada JOB 1CO

Mehrkam Energy Development Co.
179 East Road #2
Hamburg, PA 19526
(215) 562-8856
Terrance or Helena Mehrkam or Karen Votyas
EDC, Winco

M&B Alternate
Energy Systems
Dell Acres, Park A
Pieree, SD 57501

Mountain Valley Energy, Inc.
Sugarloaf Star Route
Boulder, CO 80302
(303) 442-6161
(303) 499-4494
Robert B. Keller
Enertech, Dunlite, Sencenbaugh, Wincharger, Sparco

Natural Energy Systems
Callen Avenue
Kert City, MI 49330

Natural Power Systems, Inc.
3316 Augusta Avenue
Omaha, NB 68144
(402) 334-5881
John Traudt
Dakota Wind & Sun, Dunlite, Sencenbaugh, Sparco

Ocean Wind Electric Company
29 Curtis Corner Road
Peace Dale, RI 02879

James Beech Brennan
Dakota Wind & Sun

O'Brock Windmill Sales
Route 1, 12th Street
North Benton, OH 44449
(216) 584-4681
Ken O'Brock
Baker

The Other Utility Company
Division of Environmental Energies, Inc.
P.O. Box 768
Beulah, MI
(616) 882-5095
Paul Havtlieb
Environmental Energies HWT, Elektro, Dunlite

Pacific Energy Systems
615 Romero Canyon Road
Santa Barbara, CA 93018
(805) 969-6603
Fred Carr
North Wind

The Power Company
P.O. Box 543
East Troy, WI 53120
(414) 363-3588
Anthony A. Nedd
Dakota, Dynergy, Windworks, Natural Power, Aeolian Kinetics

Prairie Sun & Wind Company
4408 - 62nd Street
Lubbock, TX 79409
(806) 795-1412
Ken Ketner
Reconditioned Older Models, Winco, Aeropower, Dakota Wind & Sun, Aermotor

Price Properties
5211 S.W. Vermont Street
Portland, OR 97219

Real Gas & Electric
P.O. Box 193
Shingletown, CA 96088
(916) 474-3852
Solomon Kagin
Elektro, Dunlite

Rede Corporation
P.O. Box 212
Providence, RI 02901
(401) 751-7333
Ronald Beckman

Sanken Wind Electric, Inc.
Kingman, AZ 86401
(602) 757-2526

Schupback, Ralph
321 - 13th Street
Alva, OK 73717
(405) 327-1685
Ralph Schupbach
Winco

Shingletown Electric
P.O. Box 237
Shingletown, CA 96088
(916) 474-3852
Robert E. Eckert
Elektro, Dunlite

Sunflower Power Company
Route 1, Box 93-A
Oskaloosa, KS 66066
(913) 597-5603

Steve Blake
Reconditioned Older Models, North Wind

Sunglow Alternative Energy Systems
8261 West State Street
Boise, ID 83702

Wind Engineering Corporation
Box 5936
Lubbock, TX 79417
(806) 763-3182
Windgen, Reconditioned Older Models, Dynergy, Dakota Wind & Sun

Wind, Wood, & Water
P.O. Box 193
Shokan, NY 12481

Windependence Electric
9080 Beeman Road
Waterloo, MI 48118
(313) 475-9669
Craig Toepfer
Reconditioned Older Models, others

Windlight Workshop
Rt. 2, Box 271
Santa Fe, NM 87501
(505) 471-2573
Mark Dankoff
Reconditioned Older Models

ANEMOMETERS AND RECORDERS

Aeolian Kinetic
P.O. Box 100
Providence, RI 02901
(401) 274-3690

Aircraft Components
North Shore Drive
Benton Harbor, MI 49022

Bendix Environmental Science Division
1400 Taylor Avenue
Baltimore, MD 21204

Climet, Inc.
1620 West Colton Avenue
Redlands, CA 92373
(714) 793-2788

Danforth
Division of the Eastern Company
Portland, ME 04103

Davis Instrument Manufacturing Co., Inc.
513 East 36th Street
Baltimore, MD 21218

Dwyer Instruments, Inc.
P.O. Box 373
Michigan City, IN 46360
(219) 872-9141
(609) 448-9200 (Hightstown, NJ)
(404) 427-9406 (Marietta, GA)
(714) 991-6720 (Anaheim, CA)

Kahl Scientific Instruments Corp.
P.O. Box 1166
El Cajon
San Diego, CA 92022
(714) 444-2158
(714) 444-5944

Kenyon Marine
P.O. Box 308
Guilford, CT 06437
(203) 453-4374

Kenyon Marine
2730B S. Main Street
Santa Ana, CA 92702
(714) 546-1101

Maximum, Inc.
42 South Avenue
Natick, MA 01760
(617) 785-0113

Meteorology Research, Inc.
P.O. Box 637
Altadena, CA 91001
(213) 791-1901

Natural Power, Inc.
New Boxton, NH 03070
(603) 487-5512

Sencenbaugh Wind Electric
P.O. Box 11174
Palo Alto, CA 94306

Sign X Laboratories, Inc.
Essex, CT 06426
(203) 767-1700

R. A. Simerl Instrument
Division
238 West Street
Annapolis, MD 21401
(301) 849-8667

M.C. Stewart Co.
Ashburnham, MA 01430

Taylor Instruments
Arden, NC 28704

Texas Electronics, Inc.
5529 Redfield Street
Box 7151, Inwood Station
Dallas, TX 75209

Thermo-Systems, Inc.
500 Cardigan Road
P.O. Box 3394
St. Paul, MN 55165
(612) 483-0900

Westberg Manufacturing Co.
3400 Westach Way
Sonoma, CA 95476
(707) 938-2121

Robert E. White
Instruments, Inc.
33 Commercial Wharf
Boston, MA 02110

WeatherMeasure Corporation
P.O. Box 41257
Sacramento, CA 95841

Windflower Wind-Computer
Lund Enterprises, Inc.
1180 Industrial Avenue
Escondido, CA 92025
(714) 746-1211

Wind Power Systems, Inc.
P.O. Box 17323
San Diego, CA 92117
(714) 452-7040

WIND MACHINE TOWERS

American Tower Company
Shelby, OH 44875

Astro Research Corporation
P.O. Box 4128
Santa Barbara, CA 93103

Bayshore Concrete
Bayonne, NJ

Natural Power, Inc.
Francestown Turnpike
New Boston, NH 03070

North Wind Power Company
Box 315
Warren, VT 05674

Solargy Corporation
17914 E. Warren Avenue
Detroit, MI 48224

Stainless, Inc.
North Wales, PA 19454

Tri-Ex Tower Corporation
7182 Rasmussen Avenue
Visalia, CA 93277

Unarco-Rohn
6718 West Plank Road
P.O. Box 2000
Peoria, IL 61601

Valmont Industries, Inc.
Valley, NB 68064

Windlight Workshop
Rt. 2, Box 271
Santa Fe, NM 87501

BATTERIES

Batteries Manufacturing Company
14694 Dequindu
Detroit, MI 48212

Bright Star
602 Getty Avenue
Clifton, NJ 07015

Burgress Division of
Clevite Corporation
Box 3140
St. Paul, MN 55101

C&D Batteries Eltuce Corp.
Washington & Chewy Streets
Conshohocken, PA 19428

Century Storage Battery
Company, LTC
Birmingham Street
Alexandria, Australia

Delatron Systems Corporation
553 Lively Boulevard
Elk Grove Village, IL 60007

Delco-Remy Division of GM
P.O. Box 2439
Anderson, IN 46011

Eggle-Pichen Industries
P.O. Box 47
Joplin, MO 64801

RCA
415 South 5th Street
Harrison, NJ 07029

Surrette Storage Battery Co., Inc.
P.O. Box 711
Salem, MA 01970

Trojan Batteries
1125 Mariposa Street
San Francisco, CA 94107

ESB Incorporated - Willard
P.O. Box 6949
Cleveland, OH 44101

Exide
5 Pen Center Plaza
Philadelphia, PA 19103

Ever Ready—Union
Carbide Corp.
270 Park Avenue
New York, NY 10017

Globe Union
5757 N. Greenbay Avenue
Milwaukee, WI 53201

Gould Incorporated
485 Calhoun Street
Trenton, NJ 08618

Gulton
212 T Dorham Avenue
Metuchen, NJ 00840

Keystone Battery Company
8301 Imperial Drive
Waco, TX 76710

Mule Battery Company
325-T Valley Street
Providence, RI 02908

INVERTERS

Allied Electronics
2400 W. Washington
Boulevard
Chicago, IL 60612
(312) 421-4200
350-1000 watt (W); 12 volt (V) input

ATR Electronics, Incorporated
300 East 4th Street
St. Paul, MN 55101
275-1100 W; 12 V input

Carter Motor Company
2711 West George Street
Chicago, IL 60618
210-740 W; variable voltage input

Dynamote Corporation
1130 N.W. 85th
Seattle, WA 98117
Automatic dc to 120 V ac, 60 Hz static inverters

Eico Electronic Instrument Co.
283 Malta Street
Brooklyn, NY 11207
110-220 W; 12-200 V input

Electro Sales Co., Inc.
100 Fellsway West
Somerville, MA 02145
(617) 666-0500
20-2000 W; 12-200 V input

Elgar Corporation
8225 Mercury Court
San Diego, CA 92111
600-1000 W; 24-48 V dc input

Best Energy Systems
for Tomorrow
P.O. Box 280
Necedah, WI 54646
(800) 356-5794
1000-5000 W; 12-48 V input

Gemini Synchronous Inverters
Windworks
Box 329, Route 3
Mukwonago, WI 53149
(414) 363-4408
4-1000 kW; variable voltage input

Heath Company
Benton Harbor, MI 49002
(616) 983-3961
175-W; 12 V input

LaMarche Manufacturing Company
106 Bradrock Drive
Des Plaines, IL 60018
(312) 279-0831
100-10,000 W; 24-120 V input

Newark Electronics
500 N. Pulaski Road
Chicago, IL
(312) 638-4411
100-250 W; 12 V input

Nova Electric
268 Hillside Avenue
Nutley, NJ 07110
(201) 661-3432

Ratelco, Incorporated
610 Pontius Avenue, N.
Seattle, WA 98109
(206) 624-7770
250-2500 W; 24-100 V input

Real Gas and Electric, Incorporated
P.O. Box F
Santa Rosa, CA 95402
5 kVa maximum ac output; 75-200 V dc input

Soleq Corporation
5969 Elston Avenue
Chicago, IL 60646
(312) 792-3811
1500-6000 W; 12-112 V input

Topaz Electronics
3855 Ruffil Road
San Diego, CA 92123
(714) 279-0831
200-3000 W; 12-125 V input

Willmore Electronics
Box 2973
Durham, NC 27705
(919) 489-3318
45-1500 W; 12-120 V input

Bibliography

This is by no means a complete list of information available on solar energy as more data becomes available every day. These are the sources I found useful in preparing this book.

Alternate Energy Sources of Hawaii. Hawaii Natural Energy Institute, University of Hawaii. February, 1975.

Braden, Spruille III. *Graphic Standards of Solar Energy*. Boston: CBI Publishing Co., 1977.

Building Blocks for Solar Energy. Rockville, Maryland: Solarex Corporation, 1979.

Ching, Francis D.K. *Building Construction Illustrated*. New York: Van Nostrand Reinhold, 1975.

Clews, Henry. *Electric Power from the Wind*. Norwich, Vermont: Henry M. Clews; 1973.

Climatic Atlas of the United States. U.S. Department of Commerce. 1974.

Coleman, Jeremy. "Experiences with Attached Greenhouses." *Solar Age*. June, 1979.

Crowther, Richard L. *Sun Earth*. Denver, Colorado: Crowther Solar Group, 1976.

Duffie, J.A. *Reflective Solar Cookie Designs*. Transactions: conference on the use of solar energy, Volume 3, part 2, pp. 79-86. Tucson, Arizona. 1955.

Evaluation of Solar Cookers. Volunteers in Technical Assistance. Mt. Rainier, Maryland.

Foster, William M. *Build-It Book of Solar Heating Projects.* Blue Ridge Summit, Pennsylvania; TAB BOOKS Inc., 1977.

Foster, William M. *Homeowners' Guide to Solar Heating and Cooling*. Blue Ridge Summit, Pennsylvania: TAB BOOKS Inc., 1976.

Foster, William M. *Sizing Solar Pool Heating Systems for Specific Geographic Conditions*. Swimming Pool Weekly, Data and Reference Annual Edition. 1978.

Fredette, Richard C. III. "Glazing Angles and Orientation." *Alternative Sources of Energy*, No. 47. January/February, 1981.

Gipe, Paul. "Wind and the Grid." *Alternative Sources of Energy*. November/December 1980.

Golding, E.W. *The Generation of Electricity by Wind Power.* New York: John Wiley & Sons. 1977.

Haupert, David; Shindelar, Carol Ann. "Add a Solar Greenhouse." *Better Homes and Gardens*. May, 1980.

Hindrichs, H.S.; Wolfert, C.K. *Wind and the Ventilation of Agricultural Buildings*. H.C. Products Company, Princeville, Illinois. 1977.

Howell, Yvonne; Adler, David. "How Silicon Solar Cells Work." *Sunworld*, Volume 4, Number 1. 1980.

Is the Wind a Practical Source of Energy for You? U.S. Department of Energy, Rocky Flats Wind Systems Program, TM-1P/80-3. September, 1980.

Kramer, Jack. *Your Homemade Greenhouse and How To Build It.* New York: Walker & Co., 1975.

Lof, George O.G. *Performance Statistics on Solar Swimming Pool Heaters.* Swimming Pool Weekly, Data and Reference Annual Edition. 1978.

Manual of Information. Hawthorne, California: Soltec International Inc.

Merchant, Michael C. *Concentrators and Solar Photovoltaics*. Sunworld, Volume 4, Number 1. 1980.

Merriam, Marshal F. "Combating Heat Losses in Swimming Pool Heating." *Solar Engineering*. June/July, 1976.

National Photovoltaic Program. U.S. Department of Energy. June, 1980.

"Photovoltaic Systems." Staff. *Machine Design.* June, 1977.

Proceedings–Solar Energy Workshop, April 5 and 6, 1976. Tucson, Arizona.

Putnam, Barbara. "Build Your Greenhouse." *Solar Age*. June, 1979.

Root, Douglass E., Jr., et. al. "Solar Piping and Connections for Swimming Pool Systems." *Solar Engineering*. November, 1979.

Simms, David. "Wiring for Your Homemade Electricity." *Alternative Sources of Energy*. November/December, 1980.

Smith, Shane. "Horticulture in an Attached Greenhouse. *Solar Age.*" June, 1979.

Solarex Guide to Solar Electricity, The. Technical Staff, Solarex Corporation. 1979.

Solar Food Drying. Hawaii Energy Extension Service. January 31, 1980.

Solar Swimming Pool Heating System Hydraulics. Application Manual 2. Solar Industries, Farmingdale, New Jersey. 1980.

Tapping the Sun. Spectrolab, Sylmar, California.

Telkes, Maria. Personal Communication. 1977.

Telkes, Maria; Andrassy, Stella. *Practical Solar Cooking Ovens*. United Nations Conference on New Sources of Energy. Princeton, New Jersey. May 17, 1961.

Telkes, Maria. *The Solar Cooking Oven*. New York University, College of Engineering Research Division. October, 1955. January, 1958.

Wegley, Harry L., et. al. *A Siting Handbook for Small Energy Conversion Systems*. U.S. Department of Energy PNL-2521, Rev. 1. March, 1980.

Wing, Charlie. *Getting Your Mass in Gear*. Alternative Sources of Energy, No. 47. January/February, 1981.

Yanda, Bill. "The Solar Efficient Greenhouse Means Growing More for Less." *Solar Engineering*. October, 1976.

Yanda, Bill. "The Solar Greenhouse Experience." *Alternative Sources of Energy*, No. 47. January/February, 1981.

Index